Recent Titles in This Series

W9-CID-436

(Continued in the back of this publication)

CONTEMPORARY MATHEMATICS

140

Geometric Analysis

Proceedings of an AMS Special Session
held October 12–13, 1991

Eric L. Grinberg
Editor

American Mathematical Society
Providence, Rhode Island

The AMS Special Session on Geometric Analysis was held at the 868th Meeting of the American Mathematical Society at Temple University, Philadelphia, Pennsylvania, on October 12–13, 1991.

1992 *Mathematics Subject Classification.* Primary 53–06, 58–06, 52–06, 44–06.

Library of Congress Cataloging-in-Publication Data

Geometric analysis: proceedings of an AMS special session held October 12–13, 1991/Eric L. Grinberg, editor
    p. cm.—(Contemporary mathematics; v. 140)
    "The AMS Special Session on Geometric Analysis was held at the 868th Meeting of the American Mathematical Society at Temple University"—T.p. verso.
    ISBN 0-8218-5153-5 (alk. paper)
    1. Global differential geometry—Congresses. 2. Geometry, Differential—Congresses. 3. Harmonic analysis—Congresses. 4. Differential equations, Partial—Congresses. I. Grinberg, Eric, 1958– . II. AMS Special Session on Geometric Analysis (1991: Temple University) III. Series: Contemporary mathematics (American Mathematical Society); v. 140.
QA670.G46 1992
516..3′62—dc20
                                         92-30474
                                                                                 CIP

This volume was printed directly from author-prepared copy.
Portions of the volume were typeset using $\mathcal{A}\mathcal{M}\mathcal{S}$-TEX,
the American Mathematical Society's TEX macro system.

10 9 8 7 6 5 4 3 2 1   97 96 95 94 93 92

Contents

Photograph Courtesy of Philadelphia Convention & Visitors Bureau

Preface

"In the preface of a book it is customary to explain the author's aim, the reasons why he wrote the book, and what he takes to be its relationship to other treatments, earlier or contemporary, of the same subject. In the case of a philosophical work, however, such an explanation seems not only superfluous but, owing to the nature of the subject matter, altogether improper and unsuited to the end in view. For what contents and tone would be appropriate for a preface to a philosophical work? "

G.W.F. HEGEL, the Preface to the *Phenomenology* (1807)

As in Hegel's work the question arises here: what is an appropriate preface for a conference proceedings volume? One could discuss the motivation for organizing the conference, or the ideas that bind the various papers together, or some combination thereof. Still, the contributions to this volume form its substance; they are autonomous entities and should stand independently. Thus I will comment briefly on the conference and let the papers speak for themselves.

Philadelphia's center city is within easy reach of many mathematical communities and, as a transportation hub, is a very convenient gathering point. Nonetheless, few meetings have been held there in recent years. The American Mathematical Society approached Shif Berhanu and asked if he would serve as local organizer for the 868th regional meeting. He agreed, and the conference was a great success. Walnut Street witnessed a great deal of activity packed into two days. Visitors enjoyed some of the local color when, on the way to the lectures they were met by the Columbus day parade marchers along Chestnut street. This time there was no mathematical float, but some numerical analysts are already planning something for next year. It seems like decades since there was an A.M.S. national meeting in Philadelphia. The success of the 868th regional meeting, and the soon to be completed Philadelphia convention center, suggest that one ought to be organized in the near future.

This volume contains papers from the special session on geometric analysis. To be precise, there is a small, but non-void, symmetric difference between the papers in the volume and the lectures presented during the conference. But this difference is bounded from above and the papers do represent accurately the spirit of the session.

What is *geometric analysis?* The term is being used with increasing frequency. The National Science Foundation has a program that goes by this name and there is a new journal that answers to it. Still, one is hard pressed to define the subject. One way to make its meaning more precise is to hold session such as this one: the material presented serves to give the name its meaning. The current volume is slated towards integral and global differential geometry, convexity, harmonic and microlocal analysis (including Radon and Fourier transforms, and wavelets) and partial differential equations. Only time will tell if these topics will ultimately comprise a part of *geometric analysis.* It is a pleasure to thank all who were involved in organizing the conference, all who provided critical help during the hectic registration and post-registration periods, the enthusiastic participants, contributors and referees, the patient and helpful publications staff at the A.M.S., and colleagues who gave their support for the project. I especially thank Shif Berhanu, Donna Harmon, and B.J. Kahn.

Eric L. Grinberg
North Philadelphia
July 27, 1992

Contemporary Mathematics
Volume **140**, 1992

On the Radon and Riesz transforms in real hyperbolic spaces

CARLOS A. BERENSTEIN AND ENRICO CASADIO TARABUSI

ABSTRACT. We investigate the Riesz transform on the real hyperbolic space $\mathbf{H}^n$, in relation to the k-dimensional Radon transform. We also give an explicit characterization of the range of the latter as the kernel of a differential operator.

1. Introduction

The k-dimensional totally geodesic Radon transform R in the real hyperbolic space $\mathbf{H}^n$, where $1 \leq k < n$, was introduced by Helgason in [**H1**]: he proved two different kinds of inversion formulas, one valid for k even [**H1**], [**H2**], the other for any k [**H3**]. This transform and its dual (or 'backprojection') R^* have recently attracted some interest due also to their applications to Electrical Impedance Tomography (EIT): for details see, e.g., [**BC1**] and the literature cited there. Based on Helgason's observation that the operator R^*R on $\mathcal{S}(\mathbf{H}^n)$ is a convolution with a radial function, a filtered backprojection inversion formula was established in [**BC1**] of the form $p(\Delta)\mathcal{S}*R^*R = I$, where p is a polynomial, Δ the Laplace-Beltrami operator, and $\mathcal{S}$ a radial integrable function (all depending on n, k). Up to a constant factor, R^*R turns out to be an instance of the Riesz operator introduced in [**H2**, §I.6.3], denoted here by J_k: it would thus seem natural that the method employed to invert R^*R be successful in inverting J_k. We shall show here that this is the case for a large class of values of the parameter k. We are indebted to R. S. Strichartz for pointing out this problem to us.

1991 *Mathematics Subject Classification.* Primary 44A12; Secondary 44A15, 51M10.

Key words and phrases. Radon transform, Riesz transform, real hyperbolic space, totally geodesic submanifolds, differential operators.

The first author was partially supported by NSF grants DMS9000619 and CDR8803012.

The second author was partially supported by the Ministero dell'Università e della Ricerca Scientifica e Tecnologica.

This paper is in final form and no version of it will be submitted for publication elsewhere

In [**BC2**] the authors exhibited intertwining operators between Euclidean and hyperbolic Radon transform, which, besides new inversion formulas for R, allow its range to be characterized, namely by suitably pulling back Helgason's moment conditions, or range-characterizing differential operators (such as those discovered by Richter [**Rr**], Gonzalez [**G**], or Kurusa [**K1**]), from $\mathbf{R}^n$ to $\mathbf{H}^n$. Here the latter kind of characterization will be discussed in greater detail, especially with regard to invariance under automorphisms of $\mathbf{H}^n$, and explicit expressions of some of the operators will be provided in the appendix.

The authors wish to thank the referee for helpful suggestions.

2. Background material

The notation is mainly taken from [**H2**], [**BC2**]. We shall use the 'conformal disk' model for $\mathbf{H}^n$, viz., the open unit ball $\mathbf{B}^n$ of $\mathbf{R}^n$ with the metric

$$ds^2 = \frac{4dx^2}{(1 - \|x\|^2)^2},$$

where $\| \cdot \|$ denotes the Euclidean norm in $\mathbf{R}^n$. Such metric is in fact conformal to the Euclidean one dx^2 and has constant curvature -1. The induced distance between $x, y \in \mathbf{H}^n$ is

$$d(x, y) = 2 \operatorname{arc\,sinh} \frac{\|x - y\|}{\sqrt{1 - \|x\|^2}\sqrt{1 - \|y\|^2}},$$

while the Euclidean norm of x can be recovered by

$$\|x\| = \tanh \frac{d(x, o)}{2}.$$

In this model, a k-dimensional totally geodesic submanifold (k-*geodesic* for short) of $\mathbf{H}^n$ is the intersection of a $(k + 1)$-plane through the origin with a spherical cap perpendicular to $\mathbf{S}^{n-1} = \partial \mathbf{B}^n$; in particular, a geodesic is an arc of circle orthogonal to $\mathbf{S}^{n-1}$.

In geodesic polar coordinates write $x \in \mathbf{H}^n$ as $x = (\omega, r)$, where $r = d(x, o)$ and $\omega \in \mathbf{S}^{n-1}$. The hyperbolic metric is then expressed by

$$ds^2 = dr^2 + \sinh^2 r\, d\omega^2,$$

where $d\omega^2$ is the usual metric in $\mathbf{S}^{n-1}$. Correspondingly, the $(n-1)$-dimensional area of a geodesic sphere of radius r is

(2.1) $A_n(r) = \Omega_n \sinh^{n-1} r,$

$$\text{where } \Omega_n = \frac{2\pi^{n/2}}{\Gamma(n/2)} \text{ is the Euclidean area of } \mathbf{S}^{n-1}.$$

The Laplace-Beltrami operator on $\mathbf{H}^n$ is

$$\Delta = \frac{(1 - \|x\|^2)^n}{4} \sum_{j=1}^{n} \frac{\partial}{\partial x_j} \left[(1 - \|x\|^2)^{2-n} \frac{\partial}{\partial x_j} \right]$$

$$= \frac{\partial^2}{\partial r^2} + (n-1) \coth r \frac{\partial}{\partial r} + \sinh^{-2} r \Delta_S,$$

where Δ_S is the Laplace-Beltrami operator on $\mathbf{S}^{n-1}$.

The space $\mathcal{D}(\mathbf{H}^n)$ denotes as usual the space of all C^∞ functions with compact support in $\mathbf{H}^n$, i.e., it coincides with the space $\mathcal{D}(\mathbf{B}^n)$ of C^∞ functions f on $\mathbf{R}^n$ whose support is in $\mathbf{B}^n$. The Schwartz space $\mathcal{S}(\mathbf{H}^n)$ of fast decreasing functions in $\mathbf{H}^n$ is the space of C^∞ functions f on $\mathbf{H}^n$ such that for any positive integers m, j we have

$$\sup_{x \in \mathbf{H}^n} |\Delta^j f(x)| e^{md(x,o)} < \infty.$$

This is equivalent to the condition that for every multiindex $\alpha \in \mathbf{N}^n$ (where $\mathbf{N}$ is the set of nonnegative integers), the function $\mathbf{B}^n \ni x \mapsto \partial^{|\alpha|} f / \partial x^\alpha(x)$ has a continuous extension to the closed ball $\overline{\mathbf{B}}^n$ which vanishes of infinite order in $\partial \mathbf{B}^n$. In other words, the space $\mathcal{S}(\mathbf{H}^n)$ coincides with the space $\mathcal{D}(\overline{\mathbf{B}}^n)$ of C^∞ functions f in $\mathbf{R}^n$ which vanish outside $\overline{\mathbf{B}}^n$. The two spaces coincide even topologically.

The family Γ of k-geodesics of $\mathbf{H}^n$ is a homogeneous space under an action of the group $SO(1,n)$ of isometries of $\mathbf{H}^n$. Each $\gamma \in \Gamma$ carries the k-dimensional area element dm_γ induced by the volume element dv in $\mathbf{H}^n$. Hence the totally geodesic k-dimensional Radon transform $R = R_{n,k}$ is defined on the space $\mathcal{S}(\mathbf{H}^n)$ by

$$(2.2) \qquad Rf(\gamma) = \int_\gamma f(x) \, dm_\gamma(x) \qquad \text{for all } \gamma \in \Gamma.$$

The family Γ_x of elements of Γ passing through a fixed point x is a homogeneous space for the isotropy group $SO(1,n)_x$ of x, which is isomorphic to $SO(n)$. Hence Γ_x carries a normalized measure $d\mu_x$ which is invariant under $SO(1,n)_x$, and is 'independent' of x in an obvious sense. For a continuous function ϕ on Γ we can define the backprojection operator R^* by

$$R^* \phi(x) = \int_{\Gamma_x} \phi(\gamma) \, d\mu_x(\gamma) = \int_{SO(n)} \phi(gh \cdot \gamma) \, dh,$$

where g is a fixed element of $SO(1,n)$ such that $g \cdot o = x$, while h runs in $SO(n)$ and dh is the normalized invariant measure in $SO(n)$. One of the uses of the backprojection operator is to find an inversion formula for the Radon transform. This is based on the fact that, denoting by $d\theta$ the area element on the geodesic sphere $S(x, r)$ of center x and radius r, we have

$$(2.3) \qquad R^* R f(x) = \int_{\mathbf{H}^n} f(y) \mathcal{R}(d(x, y)) \, dv(y)$$

$$= \int_0^\infty \mathcal{R}(r) \left[\int_{S(x,r)} f(y) \, d\theta(y) \right] dr,$$

for a function $\mathcal{R}$ on $[0, +\infty)$ (cf. [H2, Theorem I.4.5]): interpreting $\mathcal{R}$ as a radial function on $\mathbf{H}^n$ through $\mathcal{R}(x) = \mathcal{R}(d(x, o))$ (the same abuse of notation will be extended to all radial functions), we write this integral as $\mathcal{R} * f(x)$—note that the inner integral is not normalized—: in fact both $\mathcal{R}$ and f can be pulled back as functions on the group $SO(1, n)$, convolved there, and the result, pushed to $\mathbf{H}^n$ again, coincides with the middle term of (2.3).

With the notation

$$(2.4) \qquad \Gamma \begin{bmatrix} a_1 & \cdots & a_M \\ b_1 & \cdots & b_N \end{bmatrix} = \frac{\prod_{j=1}^M \Gamma(a_j)}{\prod_{\ell=1}^N \Gamma(b_\ell)}, \qquad \Gamma(a \pm b) = \Gamma(a + b) \, \Gamma(a - b),$$

where the arguments of the Euler gamma function Γ are complex numbers $\neq$ $0, -1, -2, \dots$, the function $\mathcal{R}$ turns out to be [H1]

$$\mathcal{R}(r) = \pi^{(k-n)/2} \Gamma \begin{bmatrix} n/2 \\ k/2 \end{bmatrix} \sinh^{k-n} r.$$

3. The Radon and Riesz transforms

The convolution operators $R^* R$, for varying k, may be embedded in a family of convolution operators related to the Riesz transform [Rz], [H2], [HOW]. Namely, let $\mathcal{J}_\alpha$, given by

$$\mathcal{J}_\alpha(r) = \sinh^{\alpha - n} r,$$

act as a distribution in $\mathcal{S}(\mathbf{H}^n)$, i.e., using polar coordinates as specified earlier, define $\mathcal{J}_\alpha$ by

$$\langle \mathcal{J}_\alpha, f \rangle = \int_{\mathbf{H}^n} f(x) \sinh^{\alpha - n} d(x, o) \, dx$$

$$= \int_0^\infty \int_{\mathbf{S}^{n-1}} f(r\theta) \sinh^{\alpha - n} r \, \sinh^{n-1} r \, dr \, d\theta$$

$$= \int_0^\infty M f(r) \sinh^{\alpha - 1} r \, dr, \qquad \text{where } M f(r) = \int_{\mathbf{S}^{n-1}} f(r\theta) \, d\theta$$

(the measure $d\theta$ on $\mathbf{S}^{n-1}$ is not normalized). We shall now study these operators.

For $\Re\, \alpha > 0$ the kernel $\mathcal{J}_\alpha$ corresponds to an L^1_{loc} function on $\mathbf{H}^n$. Differentiating $\langle \mathcal{J}_\alpha, f \rangle$ with respect to α one gets

$$\int_{\mathbf{H}^n} f(x) [\log \sinh d(x, o)] \sinh^{\alpha - n} d(x, o) \, dx,$$

which still has an integrable singularity for $\Re\alpha > 0$. Hence $\alpha \mapsto \mathcal{J}_\alpha$ is a holomorphic map in the region $\{\Re\alpha > 0\}$ with values in $\mathcal{S}'(\mathbf{H}^n)$ (only use that everything is uniform over compact subsets of $\mathcal{S}(\mathbf{H}^n)$).

How to extend to $\{\Re\alpha \leq 0\}$? The equation

$$(3.1) \qquad [\Delta - \beta(\beta + n - 1)]\sinh^\beta r = \beta(\beta + n - 2)\sinh^{\beta-2} r,$$

is valid a priori for $\beta \geq 2$. Let $\beta = \alpha - n + 2$: then

$$[\Delta - (\alpha + 1)(\alpha - n + 2)]\mathcal{J}_{\alpha+2} = \alpha(\alpha - n + 2)\mathcal{J}_\alpha,$$

which a priori holds only for $\alpha \geq n$. In the sense of distributions this says that

$$\langle\mathcal{J}_{\alpha+2}, [\Delta - (\alpha + 1)(\alpha - n + 2)]f\rangle = \alpha(\alpha - n + 2)\langle\mathcal{J}_\alpha, f\rangle$$

for $\Re\alpha \geq n$. Since both sides are holomorphic functions in $\{\Re\alpha > 0\}$ this identity holds true in the whole region (one could check this by hand also by taking a small ball about $x = o$ and verifying that everything is fine as long as $\Re\alpha > 0$). We want to use this equation to analytically continue $\mathcal{J}_\alpha$ to $\{\Re\alpha \leq 0\}$ as a meromorphic function. It is clear that we can define $\mathcal{J}_\alpha$ for $-2 < \Re\alpha \leq 0$ by

$$\langle\mathcal{J}_\alpha, f\rangle = \frac{\langle\mathcal{J}_{\alpha+2}, [\Delta - (\alpha + 1)(\alpha - n + 2)]f\rangle}{\alpha(\alpha - n + 2)},$$

or (in the sense of distributions)

$$(3.2) \qquad \mathcal{J}_\alpha = \frac{\Delta - (\alpha + 1)(\alpha - n + 2)}{\alpha(\alpha - n + 2)}\mathcal{J}_{\alpha+2}.$$

The numerator of the right-hand side is holomorphic for $\Re\alpha > -2$, while the denominator vanishes at $\alpha = 0$ and $\alpha = n - 2$. The second zero can occur in the strip $-2 < \Re\alpha \leq 0$ only if $n = 2$, in which case there is a possibility of a double pole. We have therefore shown that $\mathcal{J}_\alpha$ is holomorphic in $\Re\alpha > -2$ and $\alpha \neq 0$, and has a (possibly double) pole at $\alpha = 0$.

Let us now study this pole. The easiest way is to use that for $\Re\alpha > 0$ we have

$$\langle\mathcal{J}_\alpha, f\rangle = \int_0^\infty Mf(r)\sinh^{\alpha-1} r\, dr = \int_0^\infty Mf(r)\left[\frac{\sinh r}{r}\right]^{\alpha-1} r^{\alpha-1}\, dr.$$

This reduces to the study of the behavior at $\alpha = 0$ to that of $r^{\alpha-1}$, appealing to [**H2**, Remark to Theorem I.2.40] (i.e., [**Rz**, Lemma IV §35]) or directly by expanding $[\sinh r/r]^{\alpha-1}$ as

$$\left[\frac{\sinh r}{r}\right]^{\alpha-1} = \frac{r}{\sinh r} + \alpha g(r, \alpha),$$

where $g(\cdot, \alpha) \in \mathcal{S}(\mathbf{H}^n)$ is a nice function, holomorphic at $\alpha = 0$, and observing that $r^{\alpha-1}$ is, for $\gamma = \alpha - 1$, just the function r_+^γ on $\mathbf{R}$ from [GS, §I.3.1] or [H2, formula I.2.(70)], which has a pole of order 1 at $\gamma = -1$ with residue

$$\operatorname*{Res}_{\gamma=-1} r_+^\gamma = \delta_0,$$

the Dirac delta distribution at the origin of $\mathbf{R}$. Expanding

$$r_+^\gamma = \frac{1}{\gamma+1} \operatorname*{Res}_{\gamma=-1} r_+^\gamma + \operatorname*{Pf}_{\gamma=-1} r_+^\gamma + O(\gamma+1),$$

and denoting by $\langle \cdot, \cdot \rangle_{\mathbf{R}}$ the pairing on $\mathbf{R}$, the finite part at $\gamma = -1$ is given by

$$\left\langle \operatorname*{Pf}_{\gamma=-1} r_+^\gamma, \phi \right\rangle_{\mathbf{R}} = \int_0^1 \frac{\phi(r) - \phi(0)}{r}\, dr + \int_1^\infty \frac{\phi(r)}{r}\, dr;$$

in fact for $\Re\gamma > -2$ and $\gamma \neq -1$ one has (cf. [GS, §I.3.1])

$$\langle r_+^\gamma, \phi \rangle_{\mathbf{R}} = \int_0^1 r^\gamma[\phi(r) - \phi(0)]\, dr + \frac{\phi(0)}{\gamma+1} + \int_1^\infty r^\gamma \phi(r)\, dr.$$

Therefore $\mathcal{J}_\alpha$ has indeed a *simple* pole at $\alpha = 0$ (independently of n), and

$$\begin{aligned}
\langle \mathcal{J}_\alpha, f \rangle &= \int_0^\infty Mf(r)\left[\frac{r}{\sinh r} + \alpha g(r,\alpha)\right] r^{\alpha-1}\, dr \\
&= \left\langle \frac{1}{\alpha}\delta_0 + \operatorname*{Pf}_{\alpha=0} r^{\alpha-1}, Mf(r)\left[\frac{r}{\sinh r} + \alpha g(r,\alpha)\right] \right\rangle_{\mathbf{R}} + O(\alpha) \\
&= \frac{1}{\alpha}\left\langle \delta_0, Mf(r)\frac{r}{\sinh r} \right\rangle_{\mathbf{R}} \\
&\quad + \left[\langle \delta_0, Mf(r)g(r,0) \rangle_{\mathbf{R}} + \left\langle \operatorname*{Pf}_{\alpha=0} r^{\alpha-1}, Mf(r)\frac{r}{\sinh r} \right\rangle_{\mathbf{R}}\right] + O(\alpha);
\end{aligned}$$

hence

$$\left\langle \operatorname*{Res}_{\alpha=0} \mathcal{J}_\alpha, f \right\rangle = Mf(0) = \Omega_n f(o),$$

$$\left\langle \operatorname*{Pf}_{\alpha=0} \mathcal{J}_\alpha, f \right\rangle = Mf(0)g(0,0) + \int_0^1 \left[Mf(r)\frac{r}{\sinh r} - \Omega_n f(o)\right]\frac{dr}{r}$$
$$+ \int_1^\infty Mf(r)\frac{dr}{\sinh r}.$$

To find $g(r,\alpha)$ we have

$$\left[\frac{\sinh r}{r}\right]^{\alpha-1} = \frac{r}{\sinh r} e^{\alpha \log[(\sinh r)/r]} = \frac{r}{\sinh r} + \alpha\left[\frac{r}{\sinh r} \log \frac{\sinh r}{r} + O(\alpha)\right],$$

so that

$$g(r, \alpha) = \frac{r}{\sinh r} \log \frac{\sinh r}{r} + O(\alpha),$$

$$g(r, 0) = \frac{r}{\sinh r} \log \frac{\sinh r}{r},$$

$$g(0, 0) = 0.$$

Hence

$$\langle \operatorname*{Pf}_{\alpha=0} \mathcal{J}_\alpha, f \rangle = \int_0^1 \left[Mf(r) \frac{r}{\sinh r} - \Omega_n f(o) \right] \frac{dr}{r} + \int_1^\infty Mf(r) \frac{dr}{\sinh r},$$

and, denoting the distribution $\operatorname{Pf}_{\alpha=0} \mathcal{J}_\alpha$ by $\mathcal{P}$ for brevity, we have that

$$\mathcal{J}_\alpha = \frac{1}{\alpha} \Omega_n \delta_o + \mathcal{P} + O(\alpha) \qquad \text{for } \alpha \text{ near } 0$$

(here δ_o is the Dirac delta at $o \in \mathbf{H}^n$).

As a corollary we obtain that the functional equation (3.2) can now be applied to define $\mathcal{J}_\alpha$ for $-4 < \Re \alpha \leq -2$. In fact, in the right-hand side of the equation we now have a function which is holomorphic on $\{\Re(\alpha + 2) > -2\}$, except when α is -2, 0, or $n - 2$. For $\alpha = -2$ one has a simple pole for $\mathcal{J}_\alpha$ because it has a simple pole at $\alpha = 0$. The development near $\alpha + 2 = 0$ can be obtained from that at $\alpha = 0$:

$$\mathcal{J}_\alpha = \frac{\Delta - (\alpha + 1)(\alpha - n + 2)}{\alpha(\alpha - n + 2)} \left[\frac{1}{\alpha + 2} \Omega_n \delta_o + \mathcal{P} + O(\alpha + 2) \right],$$

so that

$$\operatorname*{Res}_{\alpha = -2} \mathcal{J}_\alpha = \frac{\Delta - n}{2n} \Omega_n \delta_o,$$

$$\operatorname*{Pf}_{\alpha = -2} \mathcal{J}_\alpha = \frac{(n+2)\Delta + n^2}{4n^2} \Omega_n \delta_o + \frac{\Delta - n}{2n} \mathcal{P},$$

and

$$\mathcal{J}_\alpha = \frac{1}{\alpha + 2} \frac{\Delta - n}{2n} \Omega_n \delta_o + \left[\frac{(n+2)\Delta + n^2}{4n^2} \Omega_n \delta_o + \frac{\Delta - n}{2n} \mathcal{P} \right] + O(\alpha + 2)$$

$$\text{for } \alpha \text{ near } -2.$$

Proceeding as above we can see that $\mathcal{J}_\alpha$ can be extended as a meromorphic function on the whole complex plane with simple poles exactly at $\alpha = 0, -2, -4, \ldots$, and we can compute the respective residues and finite parts. In fact, setting

$$h(\alpha) = \frac{\Delta - (\alpha + 1)(\alpha - n + 2)}{\alpha(\alpha - n + 2)},$$

from the relation

$$\mathcal{J}_\alpha = h(\alpha) \left[\frac{1}{\alpha + 2d} \operatorname*{Res}_{\alpha = -2d+2} \mathcal{J}_\alpha + \operatorname*{Pf}_{\alpha = -2d+2} \mathcal{J}_\alpha + O(\alpha + 2d) \right],$$

valid inductively for $d = 1, 2, \ldots$, one finds the recursion relations

$$\operatorname*{Res}_{\alpha=-2d} \mathcal{J}_\alpha = h(-2d) \operatorname*{Res}_{\alpha=-2d+2} \mathcal{J}_\alpha,$$

$$\operatorname*{Pf}_{\alpha=-2d} \mathcal{J}_\alpha = \frac{dh}{d\alpha}(-2d) \operatorname*{Res}_{\alpha=-2d+2} \mathcal{J}_\alpha + h(-2d) \operatorname*{Pf}_{\alpha=-2d+2} \mathcal{J}_\alpha,$$

whence

$$\operatorname*{Res}_{\alpha=-2d} \mathcal{J}_\alpha = \left[\prod_{j=1}^{d} h(-2j) \right] \Omega_n \delta_o,$$

$$\operatorname*{Pf}_{\alpha=-2d} \mathcal{J}_\alpha = \left[\sum_{i=1}^{d} \frac{dh}{d\alpha}(-2i) \prod_{\substack{1 \le j \le d \\ j \ne i}} h(-2j) \right] \Omega_n \delta_o + \left[\prod_{j=1}^{d} h(-2j) \right] \mathcal{P}.$$

In the Riesz transform proper (by analogy with [**H2**, §I.6.3], where nevertheless the integration is over Q_{-1}) one has to eliminate the factor $\beta(\beta + n - 2)$ in the functional equation (3.1). Define

$$\mathcal{I}_\alpha = \frac{1}{K_n(\alpha)} \mathcal{J}_\alpha,$$

with K_n a function to be determined. Then (3.1) becomes

$$[\Delta - (\alpha + 1)(\alpha - n + 2)]\mathcal{I}_{\alpha+2} = \frac{\alpha(\alpha - n + 2)K_n(\alpha)}{K_n(\alpha + 2)} \mathcal{I}_\alpha.$$

If we choose K_n so that

$$\frac{\alpha(\alpha - n + 2)K_n(\alpha)}{K_n(\alpha + 2)} \equiv 1,$$

then the functional equation is

$$[\Delta - (\alpha + 1)(\alpha - n + 2)]\mathcal{I}_{\alpha+2} = \mathcal{I}_\alpha,$$

where

$$K_n(\alpha) = c_n 2^\alpha \Gamma\left(\frac{\alpha}{2}\right) \Gamma\left(\frac{\alpha + 2 - n}{2}\right)$$

does the trick for any $c_n > 0$. Note now that $1/K_n(\alpha)$ vanishes at

$$\alpha \in (-2\mathbf{N}) \cup (-2\mathbf{N} + n - 2) = \{0, -2, -4, \ldots\} \cup \{n - 2, n - 4, \ldots\}.$$

Hence the $\mathcal{S}'$-valued function $\mathcal{I}_\alpha$ is entire. But if n is even then $\mathcal{I}_\alpha$ is zero at every point of $-2\mathbf{N}$. If n is odd then it will give zero values at all odd integers $\le n - 2$, though the original transform $\mathcal{J}_\alpha$ did not. For n odd we have

$$\mathcal{I}_0 = \frac{\Omega_n \frac{1}{2} \frac{d(1/\Gamma)}{d\alpha}(0)}{c_n \Gamma(1 - n/2)} \delta_o.$$

Now

$$\frac{d(1/\Gamma)}{d\alpha}(0) = 1, \qquad \Gamma(1 - n/2)\,\Gamma(n/2) = \frac{\pi}{\sin(\pi n/2)}, \qquad \Omega_n = \frac{2\pi^{n/2}}{\Gamma(n/2)},$$

therefore

$$\mathcal{I}_0 = c_n^{-1}\pi^{n/2-1}\sin(n\pi/2)\,\delta_o.$$

Setting

$$c_n = (-1)^{(n-1)/2}\pi^{n/2-1},$$

and defining $I^\alpha f = \mathcal{I}_\alpha * f$, we finally obtain that

$$I^0 f = f \qquad \text{for } n \text{ odd.}$$

(Questions of analytic continuation of the kind just described have been thoroughly investigated by Horváth, Ortner, and Wagner in their work on hypersingular integrals: see, e.g., [HOW], [W].)

For our purposes it is therefore better to keep track of the meromorphic operator-valued function $J_\alpha f = \mathcal{J}_\alpha * f$, acting in the space $\mathcal{S}(\mathbf{H}^n)$. The convolution operator R^*R corresponds to $\alpha = k$, more precisely

$$R^*R = \pi^{(k-n)/2}\Gamma\begin{bmatrix} n/2 \\ k/2 \end{bmatrix} J_k.$$

In [BC1] the authors investigated R^*R using symbolic calculus. The main difficulty was the explicit identification of the inverse operator. We shall show here that the techniques used for that paper can be generalized to invert some of the operators J_α. For the benefit of the reader we recall in the sequel the inversion formulas obtained for R^*R. The inversion is achieved by means of a radial distribution $\mathcal{T}$ such that

$$\mathcal{T} * \mathcal{R} = \int_{\mathbf{H}^n} \mathcal{T}(d(y,o))\,\mathcal{R}(d(\,\cdot\,,y))\,dv(y) = \delta_o.$$

That is, for any even function $\chi \in \mathcal{D}(\mathbf{R})$ one should have

$$\int_{\mathbf{H}^n}\int_{\mathbf{H}^n} \mathcal{T}(d(y,o))\,\mathcal{R}(d(x,y))\,\chi(d(x,o))\,dv(x)\,dv(y) = \chi(o)$$

(these integrations are intended in the distribution sense if $\mathcal{T}$ is not a function). The inversion formula is thus provided by the convolution with an $SO(n)$-bi-invariant distribution $\mathcal{T}$ in $SO(1,n)$. Specifically, such a distribution is of the form $\mathcal{T} = p(\Delta)\mathcal{S}$, where p is a polynomial and $\mathcal{S}$ is a radial function (sufficiently integrable that the inversion formula can be applied to functions in $\mathcal{D}(\mathbf{H}^n)$).

The spherical Fourier transform $\hat{f}$ of a radial function (or distribution) f is defined by

$$(3.3) \qquad \hat{f}(\lambda) = \int_{\mathbf{H}^n} f(x)\,\phi_{-\lambda}(x)\,dv(x) = \int_0^\infty f(r)\,\phi_{-\lambda}(r)\,A_n(r)\,dr,$$

where $A_n(r)$ is given in (2.1) and $\phi_\lambda = \phi_{-\lambda}$, called *spherical function*, is the radial eigenfunction of Δ of eigenvalue $-(n-1)^2/4 - \lambda^2$ such that $\phi_\lambda(o) = 1$: it is given by

$$(3.4) \qquad \phi_\lambda(r) = 2^{n/2-1}\Gamma(n/2)\sinh^{1-n/2} r P_{i\lambda-1/2}^{1-n/2}(\cosh r), \qquad \text{for } r \geq 0,$$

where $P_\nu^\mu = P_{-\nu-1}^\mu$ is an associated Legendre function (see, e.g., [BZ]). The asymptotic behavior of P_ν^μ is (cf. [GR, 8.771.1, 776.1])

$$(3.5) \qquad P_\nu^\mu(u) = \begin{cases} O((u-1)^{-\Re\mu/2}) & \text{as } u \to 1^+, \\ O(u^{\max\{\Re\nu,-1-\Re\nu\}}) & \text{as } u \to +\infty. \end{cases}$$

The spherical Fourier transform is inverted by

$$f(z) = c_n \int_{\mathbf{R}} \hat{f}(\lambda)\,\phi_\lambda(z)\,|\mathbf{c}(\lambda)|^{-2}\,d\lambda,$$

where c_n is a suitable constant and

$$\mathbf{c}(\lambda) = 2^{n/2-1}\pi^{-1/2}\Gamma\begin{bmatrix} 3(n-1)/2 & i\lambda \\ n-1 & (n-1)/2 + i\lambda \end{bmatrix}$$

is Harish-Chandra's c-function. The spherical Fourier transform and its inversion are valid for radial functions in $L_c^2(\mathbf{H}^n, dv)$, but can be extended to spaces of distributions. Usually $\hat{f}$ is called the *symbol* of the operator of convolution in $\mathbf{H}^n$ with the radial distribution f. For instance $\hat{\delta} \equiv 1$ and $\hat{\Delta}(\lambda) = -(n-1)^2/4 - \lambda^2$, by the definition of ϕ_λ; more generally, if p is a polynomial we have $p(\Delta)^\wedge(\lambda) = p(-(n-1)^2/4 - \lambda^2)$. Furthermore $(f * g)^\wedge(\lambda) = \hat{f}(\lambda)\hat{g}(\lambda)$ for f, g radial; hence

$$(p(\Delta)f)^\wedge(\lambda) = p(-(n-1)^2/4 - \lambda^2)\,\hat{f}(\lambda).$$

The reader is referred to [H2] for the above assertions.

In view of (3.3), we say that a radial function f is of *symbol class* if $f\phi_{-\lambda} \in L^1(\mathbf{H}^n, dv)$ for every $\lambda \in \mathbf{R}$—in this case f is completely determined by the symbol $\hat{f}$ of the operator of convolution with it. Integrating in polar coordinates and recalling (3.4), (3.5), if f is continuous on $(0, +\infty)$ this condition can be rephrased as

$$\int_0^1 |f(r)|r^{n-1}\,dr + \int_1^\infty |f(r)|e^{(n-1)r/2}\,dr < \infty.$$

Let d_0, d_1 be the smallest nonnegative integers such that $(n+1)/2 + 2d_0$ and $(n-1)/2 + 2d_1$ are greater than k, i.e.,

$$d_0 = \max\{0, \lfloor k/2 - (n-3)/4 \rfloor\},$$
$$d_1 = \max\{0, \lfloor k/2 - (n-5)/4 \rfloor\}$$

(the 'floor function' $\lfloor t \rfloor$ denotes the integral part of $t \in \mathbf{R}$): observe that d_1 equals either d_0 or $d_0 + 1$. Consider the following polynomials:

$$p(t) = \prod_{j=1}^{k} [-t - (n-k-2+j)(k+1-j)],$$

$$p_0(t) = \prod_{\text{even } j=2}^{2d_0} [-t - (n-k-2+j)(k+1-j)],$$

$$p_1(t) = \prod_{\text{odd } j=1}^{2d_1-1} [-t - (n-k-2+j)(k+1-j)],$$

$$p_2(t) = \prod_{j=d_0+d_1+1}^{k} [-t - (n-k-2+j)(k+1-j)]$$

(as usual, empty products are made equal to 1): notice that $p = p_0 p_1 p_2$, and that p_0, p_1 are of degrees d_0, d_1, respectively.

In [**BC1**] it is shown that if S is the operator of convolution associated to the radial function

$$S(r) = \sinh^{k-n} r \cosh r,$$

then $p(\Delta) S R^* R = I$, which is the filtered backprojection formula mentioned in the introduction. The proof is achieved by multiplying the relevant symbols. The inversion formula above is also valid for the remaining cases $(2, 1)$ and $(3, 1)$, but in both of them the kernel S needs to be regularized (cf. [**BC1**, Theorems 4.3, 4.4]): in the former case, for instance, we need to replace $S = \coth r$ with $\tilde{S} = \coth r - 1$. Other inversion formulas, which do not factor through $R^* R$ can be found in [**H3**].

Analogous to [**BC1**, Lemma 3.1] we now have

LEMMA 3.1. *The function $\mathcal{J}_\alpha$ is of symbol class if $0 < \Re e\, \alpha < (n+1)/2$.*   □

It follows from [**GR**, 7.151.1] that for α in the strip of the above lemma we have, setting $\nu = i\lambda - 1/2$ and using notation (2.4),

(3.6)
$$\hat{\mathcal{J}}_\alpha(\lambda) = \int_0^\infty \mathcal{J}_\alpha(r)\, \phi_{-\lambda}(r)\, A_n(r)\, dr$$

$$= (2\pi)^{n/2} \int_0^\infty \sinh^{\alpha-n/2} r\, P_\nu^{1-n/2}(\cosh r)\, dr$$

$$= \pi^{n/2} \Gamma \begin{bmatrix} \alpha/2 & (n+1-2\alpha)/4 \pm i\lambda/2 \\ (n-\alpha)/2 & (n+1)/4 \pm i\lambda/2 \end{bmatrix},$$

which is a priori valid for

$$0 < \Re e\, \alpha < (n+1)/2 - \Im m\, \lambda;$$

assuming $\lambda \in \mathbf{R}$—as is usually the case—the above expression of the symbol of $\mathcal{J}_\alpha$ is true for

$$0 < \Re e\, \alpha < (n+1)/2.$$

Let $\sigma(\alpha, \lambda)$ equal the bottom line of (3.6) for all the values of α, λ for which it makes sense. Then the function $\alpha \mapsto \sigma(\alpha, \lambda)$ is meromorphic, it being a quotient of meromorphic functions. On the other hand, from the functional equation and regularization we can see that the function $\alpha \mapsto \hat{\mathcal{J}}_\alpha(\lambda)$ also can be extended as a meromorphic function beyond the range $0 < \Re\alpha < (n+1)/2$, where $\mathcal{J}_\alpha$ is of symbol class. There are two ways of doing this. The easiest method is to observe that as a consequence of [H2, Corollary IV.8.2] (compare with (3.4), (3.5)) the functions in $\hat{\mathcal{S}}(\mathbf{H}^n)$ are holomorphic in $\mathbf{R}+i(-n-1, n+1)$ and have fast decrease (see [H2, Exercise IV.C.6]). Then $\mathcal{S}'(\mathbf{H}^n)\hat{\ }$ corresponds to the dual of $\hat{\mathcal{S}}(\mathbf{H}^n)$, which contains $\mathcal{S}_{\mathrm{hol}}(\mathbf{R})$.

Since $\alpha \mapsto \mathcal{J}_\alpha$ is meromorphic into $\mathcal{S}'(\mathbf{H}^n)$ then $\alpha \mapsto \hat{\mathcal{J}}_\alpha$ is meromorphic into $\mathcal{S}'(\mathbf{H}^n)\hat{\ }$. But for each α (except the poles, to be discussed momentarily) the function $\sigma(\alpha, \lambda)$ has at most polynomial growth in the real λ axis: this is easily verified from (3.6) making use of Stirling's formula [GR, 8.328.1]. The only problem is that for some complex values of α one can have simple poles in the real λ axis—namely when

$$(n + 1 - 2\alpha)/4 \pm i\lambda/2 \in \mathbf{N},$$

that is,

$$\lambda = \pm\,\Im\mathfrak{m}\,\alpha \qquad \text{and} \qquad \Re\alpha \in 2\mathbf{N} + (n+1)/2.$$

In particular, this can only occur when α is such that $\Re\alpha \in 2\mathbf{N}+(n+1)/2$. Note that this locus is disjoint from $-2\mathbf{N}$, where $\sigma(\alpha, \lambda)$ has a pole independently of λ (recall that $-2\mathbf{N}$ is also the set of poles for $\mathcal{S}_\alpha$ itself). These poles of λ create no problem, since $\sigma(\alpha, \cdot)$ acts on $\hat{\mathcal{S}}(\mathbf{H}^n)$ by integration, i.e., $\int_{-\infty}^\infty \sigma(\alpha, \lambda)\,\phi(\lambda)\,d\lambda$, and the integral can be defined by deforming the path of integration a bit around the pole—this amounts to computing the Cauchy principal value of the integral about the pole λ_0:

$$\lim_{\epsilon\to 0}\left[\int_{-\infty}^{\lambda_0-\epsilon} + \int_{\lambda_0+\epsilon}^\infty\right].$$

It is also clear that the map $\alpha \mapsto \sigma(\alpha, \cdot)$ is meromorphic into $\hat{\mathcal{S}}'(\mathbf{H}^n)$. Since it coincides with $\hat{\mathcal{J}}_\alpha$ for $0 < \Re\alpha < (n+1)/2$ we have equality everywhere. One can obtain the same result by extending $\hat{\mathcal{J}}_\alpha$ using the functional relation and the fact that $\hat{\Delta}(\lambda) = -(n-1)^2/4 - \lambda^2$, as done in [BC1]. At the poles $\alpha \in -2\mathbf{N}$ we can compute the residue and finite part of σ either directly or using the previous development for $\mathcal{J}_\alpha$. For instance

$$\operatorname*{Res}_{\alpha=0} \hat{\mathcal{J}}_\alpha = \Omega_n\hat{\delta}_0 \equiv \Omega_n.$$

The same argument also shows that the functions

$$\mathcal{J}_{\alpha,1} = \sinh^{\alpha-n} r \cosh r,$$

and, more generally,

$$\mathcal{J}_{\alpha,\beta} = \sinh^{\alpha-n} r \cosh^\beta r,$$

which are locally integrable for $\Re\,\alpha > 0$ and β arbitrary, define meromorphic maps

$$\alpha \mapsto \mathcal{J}_{\alpha,1},$$
$$(\alpha,\beta) \mapsto \mathcal{J}_{\alpha,\beta}$$

into $\mathcal{S}'(\mathbf{H}^n)$, with possible poles at $\alpha \in -2\mathbf{N}$ and $(\alpha,\beta) \in -2\mathbf{N}\times\mathbf{C}$, respectively. All we have to do is to observe that the map of multiplication by $\cosh^{\beta} r$ is an isomorphism of $\mathcal{S}(\mathbf{H}^n)$ into itself. The argument used for proving [BC1, Lemma 4.1] can be carried out when $0 < \Re\,\alpha < (n-1)/2$ and leads to

$$\hat{\mathcal{J}}_{\alpha,1}(\lambda) = \pi^{n/2}\Gamma\left[\begin{array}{cc} \alpha/2 & (n-1-2\alpha)/4\pm i\lambda/2 \\ (n-\alpha)/2 & (n-1)/4\pm i\lambda/2 \end{array}\right].$$

The earlier considerations show that, correctly interpreted, this identity of distributions in $\hat{\mathcal{S}}(\mathbf{H}^n)'$ holds for every α.

It is now easy to extend the validity of the inversion formula of [BC1] recalled above to any positive integer $\alpha = k$ (i.e., k is not constrained to be smaller than n any more)—however, the special cases to be treated separately become infinitely many. Consistently with the previous notation, set $J_{\alpha,\beta}f = \mathcal{J}_{\alpha,\beta} * f$ for $f \in \mathcal{S}(\mathbf{H}^n)$.

THEOREM 3.2.   *The kernels $p_0(\Delta)\mathcal{J}_k$ and $p_1(\Delta)\mathcal{J}_{k,1}$ are of symbol class unless $n = 2, 3$ (any k), or $n = 4, 5$ and k is even. In general, unless $n = 4, 5$ and k is odd we have the identity (in the sense of radial distributions)*

$$p(\Delta)(\mathcal{J}_{k,1} * \mathcal{J}_k) = p_2(\Delta)(p_1(\Delta)\mathcal{J}_{k,1} * p_0(\Delta)\mathcal{J}_k),$$

and the transform J_k on the space $\mathcal{D}(\mathbf{H}^n)$ is inverted by

$$p(\Delta)J_{k,1}J_k = c_{n,k}I, \qquad \text{with } c_{n,k} = 4^k\pi^n\Gamma\left[\begin{array}{c} k/2 \\ (n-k)/2 \end{array}\right]^2.$$

In the remaining cases the regularization of the kernel $\mathcal{J}_{k,1}$ for k odd is reconducted to the case $k = 1$ (established in [BC1, Theorems 4.3, 4.4]): if $n = 2$, for instance, replace $\mathcal{J}_{k,1}$ with

$$\tilde{\mathcal{J}}_{k,1} = \mathcal{J}_{k,1} - \left[\frac{(k-2)!!}{(k-1)!!}\right]^2 \mathcal{J}_{k+1}$$

(set $(-1)!! = 1$), and the degree d_1 with $\tilde{d}_1 = d_1 - 1 = (k-1)/2$, so that the polynomial p_1 changes to $\tilde{p}_1$ accordingly, and we have

$$\tilde{p}_1(\Delta)\tilde{\mathcal{J}}_{k,1} = (-1)^{(k-1)/2}(k-2)!!^2\,\mathcal{J}_{1,1},$$

which is the regularized kernel of [BC1, Theorem 4.3]. □

The question remains open of the inversion of $\mathcal{J}_\alpha$, or in general $\mathcal{J}_{\alpha,\beta}$, for the remaining values of α, β.

4. Characterizations of the range of the Radon transform

In [BC2] we showed that the Radon transform on $\mathbf{H}^n$ intertwines with that in the Euclidean space $\mathbf{R}^n$, and obtained a characterization of the range using Helgason's moment conditions. Such intertwining—independently found by Kurusa [K2] who uses it to prove support theorems—is exploited here to characterize the range as the kernel of differential operators. To this end let us briefly recall the definitions of the intertwining operators as given in [BC2].

Let $\Xi = \Xi_{n,k}$ be the set of intersections of $(n-k)$-dimensional subspaces of $\mathbf{R}^n$ with the unit sphere $\mathbf{S}^{n-1}$. Denote by G the Grassmannian space of k-planes in $\mathbf{R}^n$, and by G_B the subspace of k-planes that intersect $\mathbf{B}^n$. Naturally the closure $\overline{G}_B$ of G_B is the compact subset of G consisting of k-planes intersecting $\overline{\mathbf{B}}^n$. Take polar coordinates in $\mathbf{R}^n$, so that a point x is given by a pair $(w,s) \in \mathbf{S}^{n-1} \times [0,\infty)$. Then identify G with the set of triples $(\xi,w,s) \in \Xi \times \mathbf{S}^{n-1} \times [0,\infty)$ such that $w \in \xi$: such a triple represents the k-plane which is orthogonal at the point (w,s) to the $(n-k)$-plane through the origin o generated by ξ (observe that (w,s) is the closest point of the k-plane to o). In particular $G_B = \{(\xi,w,s) \in G : s < 1\}$. In a completely analogous way Γ can be parametrized by the same set of triples, and the $(n-k)$-plane generated by ξ in $\mathbf{R}^n$ projects to an $(n-k)$-geodesic through o via the exponential map at o. In both cases the ξ coordinate is redundant if $k = n-1$ (it must be $\xi = \{\pm w\}$).

Consider the diffeomorphism $\tau\colon \mathbf{H}^n \to \mathbf{B}^n$ given in polar coordinates by

$$\tau(w,r) = (w, \tanh r) \qquad \text{for all } (w,r) \in \mathbf{H}^n.$$

As shown in [BC2, Proposition 3.1], τ maps each k-geodesic of $\mathbf{H}^n$ to the intersection with $\mathbf{B}^n$ of a k-plane of $\mathbf{R}^n$. As a consequence there is a naturally induced diffeomorphism $\tau_k\colon \Gamma \to G_B$, given in the above described coordinates by

$$\tau_k(\xi,w,r) = (\xi, w, \tanh r) \qquad \text{for all } (\xi,w,r) \in \Gamma.$$

Let ρ, σ be the functions on $\mathbf{H}^n$, Γ respectively, defined by

$$\rho(w,r) = \cosh^{k+1} r,$$
$$\sigma(\xi,w,r) = \cosh r.$$

The intertwining operators between the hyperbolic Radon transform R and its Euclidean counterpart R_E, similarly defined on $\mathcal{S}(\mathbf{R}^n)$ by

$$R_E f(\pi) = \int_\pi f(y)\, dm_\pi(y) \qquad \text{for all } \pi \in G,$$

where dm_π is the k-dimensional area element induced on π by the ordinary volume element in $\mathbf{R}^n$ (cf. (2.2)), are $\Phi\colon \mathcal{S}(\mathbf{H}^n) \to \mathcal{D}(\overline{\mathbf{B}}^n)$ and $\Psi\colon \mathcal{S}(\Gamma) \to \mathcal{D}(\overline{G}_B)$ given by

$$\Phi(f) = (\rho f) \circ \tau^{-1},$$
$$\Psi(\phi) = (\sigma\phi) \circ \tau_k^{-1}.$$

PROPOSITION 4.1. (cf. [**BC2**, Propositions 3.4, 3.5])   *The operators* Φ, Ψ *are topological isomorphisms which render commutative the diagram*

$$\begin{array}{ccc} \mathcal{S}(\mathbf{H}^n) & \xrightarrow{\;R\;} & \mathcal{S}(\Gamma) \\ \Phi\downarrow & & \Psi\downarrow \\ \mathcal{D}(\overline{\mathbf{B}}^n) & \xrightarrow{\;R_E\;} & \mathcal{D}(\overline{G}_B) \end{array}$$

i.e., $\Psi R = R_E \Phi$. *The same holds if* $\mathcal{S}(\mathbf{H}^n)$, $\mathcal{S}(\Gamma)$ *are replaced with* $\mathcal{D}(\mathbf{H}^n)$, $\mathcal{D}(\Gamma)$, *and* $\mathcal{D}(\overline{\mathbf{B}}^n)$, $\mathcal{D}(\overline{G}_B)$ *by* $\mathcal{D}(\mathbf{B}^n)$, $\mathcal{D}(G_B)$, *respectively.*

PROOF. Fix $\gamma = (\xi, \omega', r') \in \Gamma$, and set $\pi = \tau_k(\gamma) = (\xi, \omega', s')$, where $s' = \tanh r'$. If $d\omega$ is the k-dimensional measure on $\{\omega \in \mathbf{S}^{n-1} : (\omega, r) \in \gamma\}$, for $x = (\omega, r) \in \gamma$ and $y = \tau(x) = (\omega, s)$, where $s = \tanh r$, we have (cf. [**BC2**, Lemma 2.1])

$$dm_\gamma(x) = \frac{\sinh^{k+1} r}{\sinh r'} d\omega, \qquad dm_\pi(y) = \frac{s^{k+1}}{s'} d\omega,$$

so that the jacobian of τ along γ at x is

$$\frac{dm_\pi(y)}{dm_\gamma(x)} = \frac{\cosh r'}{\cosh^{k+1} r} = \frac{\sigma(\gamma)}{\rho(x)}.$$

Therefore if f is in $\mathcal{S}(\mathbf{H}^n)$ or $\mathcal{D}(\mathbf{H}^n)$ then

$$\Psi Rf(\pi) = \sigma(\gamma) \int_\gamma f \, dm_\gamma = \int_\pi (\rho f) \circ \tau^{-1} \, dm_\pi = R_E \Phi f(\pi).$$

Finally, it is easy to see that if f is fast decreasing on $\mathbf{H}^n$ then Φf extends to 0 outside $\mathbf{B}^n$ as a C^∞ function, so that Φ maps indeed $\mathcal{S}(\mathbf{H}^n)$ into $\mathcal{D}(\overline{\mathbf{B}}^n)$. □

If $1 \le k < n - 1$ the range of the Euclidean Radon transform for $\mathcal{S}(\mathbf{R}^n)$ or $\mathcal{D}(\mathbf{R}^n)$ has been characterized as the set of solutions of a homogeneous system of linear differential equations. John [**J**] first showed that the ultrahyperbolic operator in $\mathbf{R}^4$ could be used to characterize the range of the X-ray transform (i.e., $k = 1$) in $\mathbf{R}^3$. After other results by Grinberg and other authors, Richter [**Rr**] exhibited a system of translation invariant, globally defined second-order differential operators which characterize $R(\mathcal{S}(\mathbf{R}^n))$ for each $k < n-1$. Recently, other characterizations were provided by Kurusa [**K1**] with a system of ultrahyperbolic equations, and by Gonzalez [**G**] with a single fourth-order differential operator, invariant by all isometries of $\mathbf{R}^n$. To discuss the corresponding results in $\mathcal{S}(\mathbf{H}^n)$ (and $\mathcal{D}(\mathbf{H}^n)$), we shall concentrate on Richter's and Gonzalez' operators and explain how they give rise to (second-, respectively) fourth-order equations characterizing the ranges of the hyperbolic Radon transform, although the same procedure can be applied to other differential equations, such as Kurusa's.

Let $E = E_n$ be the Lie group of isometries of $\mathbf{R}^n$, let $\mathfrak{e}$ be its Lie algebra, and let $\mathfrak{A}$ be the universal enveloping algebra of $\mathfrak{e}$. Denote by β the left action of E on G. We say that there is a *germ (at the identity) of left action* β of E

on the open set G_B of G, by restriction: for every $\pi \in G_B$, only those $b \in E$ are allowed such that $\beta(b)\pi \in G_B$—these form an open set of E that depends on π and contains the identity. Denote by ν the left regular representation of E on $\mathcal{S}(G)$, so that $\nu(b)\psi = \psi \circ \beta(b)^{-1}$ for every $\psi \in \mathcal{S}(G)$, and extend to $\mathfrak{A}$ the infinitesimal left regular representation $d\nu$ on $\mathcal{S}(G)$ in the usual way: thus

$$d\nu(V_1 \cdots V_m)\psi(\pi)$$
$$= \frac{\partial^m}{\partial t_1 \cdots \partial t_m}[\nu(\exp t_1 V_1 \cdots \exp t_m V_m)\psi(\pi)]_{t_1=\cdots=t_m=0}$$
$$= \frac{\partial^m}{\partial t_1 \cdots \partial t_m}[\psi(\beta(\exp(-t_m V_m) \cdots \exp(-t_1 V_1))\pi)]_{t_1=\cdots=t_m=0}$$

$$\text{for each } V_1, \ldots, V_m \in \mathfrak{e}, \ \psi \in \mathcal{S}(G), \ \pi \in G.$$

The expression after the last equality sign can be used to define the 'infinitesimal left regular representation' $d\nu$ on $\mathcal{D}(G_B)$ (even though the 'finite' left regular representation ν on $\mathcal{D}(G_B)$ does not itself make sense). In fact, on each fixed $\pi \in G_B$ the element $\exp(-t_m V_m) \cdots \exp(-t_1 V_1)$ of E can act whenever $-a < t_1, \ldots, t_m < a$, for some positive a depending on π, $V_1, \ldots, V_m$.

Conjugating β by τ_k, we get a germ of left action α of E on Γ: that is, $\alpha(b) = \tau_k^{-1} \circ \beta(b) \circ \tau_k$. Notice that $\alpha(b)$ is an automorphism of $\mathbf{H}^n$ if and only if it keeps o fixed. (The other elements of E cannot act, through α, as hyperbolic isometries, since they are only defined on part of Γ, and push the remaining k-geodesics 'off' $\mathbf{H}^n$.) By analogy to the above situation, define the infinitesimal left regular representation $d\mu$ on $\mathcal{S}(\mathbf{H}^n)$ by

$$d\mu(V_1 \cdots V_m)\phi(\gamma)$$
$$= \frac{\partial^m}{\partial t_1 \cdots \partial t_m}[\phi(\alpha(\exp(-t_m V_m) \cdots \exp(-t_1 V_1))\gamma)]_{t_1=\cdots=t_m=0}$$

$$\text{for each } V_1, \ldots, V_m \in \mathfrak{e}, \ \phi \in \mathcal{S}(\Gamma), \ \gamma \in \Gamma.$$

Of course $d\mu(V_1 \cdots V_m)\phi = [d\nu(V_1 \cdots V_m)(\phi \circ \tau_k^{-1})] \circ \tau_k$.

Let $T_j \in \mathfrak{e}$ be the 'infinitesimal translation in the j-th coordinate' of $\mathbf{R}^n$, for $j = 1, \ldots, n$: identifying $\mathfrak{e}$ with the Lie algebra of matrices $\begin{pmatrix} L & v \\ 0 & 0 \end{pmatrix}$, where $L \in so(n)$ and $v \in \mathbf{R}^n$, then T_j corresponds to the matrix all of whose entries all vanish except the $(j, n+1)$, which equals 1. Also, let X_{ij} be the 'infinitesimal rotation around the origin in the plane of the i-th and j-th coordinates', for $i, j = 1, \ldots, n$ with $i \neq j$: the only nonzero entries of its matrix are the (i, j), which equals 1, and the (j, i), which equals -1. Richter's 'pre-operators' $V_{ijl} \in \mathfrak{A}$, for distinct $i, j, k = 1, \ldots, n$, are given by [**Rr**]

$$V_{ijl} = T_i X_{jl} + T_j X_{li} + T_l X_{ij},$$

whereas Gonzalez' is [**G**]

$$V = \sum_{1 \leq i < j < l \leq n} V_{ijl}^2.$$

Richter's differential operators $d\nu(V_{ijl})$ on $\mathcal{S}(G)$ are invariant by translations, but not by rotations, while Gonzalez' operator $d\nu(V)$ enjoys both invariances. Notice that V_{ijl} and V are independent of k, whereas the differential operators proper are not, as they act on functions defined on different Grassmannian manifolds. Stipulating that a differential operator D on $\mathcal{S}(\overline{G}_B)$ is said to be invariant by a motion $b \in E$, such that $\beta(b)G_B$ intersects G_B, if

$$D[\psi \circ \beta(b)] \equiv (D\psi) \circ \beta(b) \quad \text{on } \{\pi \in G_B : \beta(b)\pi \in G_B\} \text{ for every } \psi \in \mathcal{S}(\overline{G}_B)$$

(the equality would mean that D commutes with $\nu(b)$, if the latter made sense), the second-order operators $D_{ijl} = d\nu(V_{ijl})$ on $\mathcal{S}(\overline{G}_B)$ are, as well, invariant by translations but not by rotations, while the fourth-order $D = d\nu(V)$ is fully invariant under E.

Pulling back with Ψ, it is easy to see that the differential operators $C_{ijl} = \Psi^{-1}D_{ijl}\Psi$ on $\mathcal{S}(\Gamma)$ are given by

$$(4.1) \qquad C_{ijl}\phi = \sigma^{-1}d\mu(V_{ijl})(\sigma\phi) \qquad \text{for every } \phi \in \mathcal{S}(\Gamma).$$

It is easy to realize that C_{ijl} are neither invariant by rotations around o (because D_{ijl} are not), nor by other automorphisms of $\mathbf{H}^n$; yet, they are invariant under the germ of action α of the subgroup of E of Euclidean translations in $\mathbf{R}^n$.

Analogously Ψ pulls back D to the fourth-order operator $C = \Psi^{-1}D\Psi$ on $\mathcal{S}(\Gamma)$, given by an expression similar to (4.1). Now, C is fully invariant under the germ of action α of E, whereas the only hyperbolic isometries which leave it invariant are those obtainable as α of some element of E, i.e., the rotations around the origin. The problem is open on whether a range-characterizing operator exists that is invariant by all hyperbolic isometries.

The explicit computation of C_{ijl} and C is relegated to the appendix. We summarize the above remarks in the following statement.

THEOREM 4.2. *If $1 \le k < n - 1$, a function ϕ in $\mathcal{S}(\Gamma)$ (or $\mathcal{D}(\Gamma)$) is in the range $RS(\mathbf{H}^n)$ (respectively $RD(\mathbf{H}^n)$) of the k-dimensional hyperbolic Radon transform if and only if it is annihilated by the second-order differential operator C_{ijl} for all $1 \le i < j < l \le n$; equivalently, if and only if it is annihilated by the fourth-order differential operator C.* $\square$

Appendix: Computation
of range-characterizing differential operators

Euclidean space. For distinct indices a, b, c and distinct i, j, l define the map $\zeta = \zeta_{(a,b,c,i,j,l)} : \mathbf{R}^4 \to E$ by

$$\zeta(t_1, t_2, t_3, t_4) = \exp(-t_4 X_{ab})\exp(-t_3 T_c)\exp(-t_2 X_{ij})\exp(-t_1 T_l)$$

for $t_1, t_2, t_3, t_4 \in \mathbf{R}$. Let $\{e_1, \ldots, e_n\}$ be the standard basis of $\mathbf{R}^n$. Let $t \mapsto r_{ij}(t)$ be the one-parameter subgroup in E generated by $-X_{ij}$: so $r_{ij}(t) = \exp(-tX_{ij}) = r_{ji}(t)^{-1}$ is the rotation of angle t in the (i, j) plane. Such subgroup has natural actions on Ξ and on $\mathbf{S}^{n-1}$, which will merely be denoted by juxtaposition. Let P_ξ be the orthogonal projection of $\mathbf{R}^n$ onto $\xi \in \Xi$. For every k-plane $\pi = (\xi, \omega, s) \in G$ and for t_1, t_2, t_3, t_4 small in absolute value we have

$$\zeta(t_1, t_2, t_3, t_4)(\pi) = (r_{ab}(t_4) r_{ij}(t_2)\xi,$$
$$r_{ab}(t_4) r_{ij}(t_2) \operatorname{dir}(s\omega - t_1 P_\xi e_l - t_3 P_\xi r_{ji}(t_2) e_c),$$
$$\|s\omega - t_1 P_\xi e_l - t_3 P_\xi r_{ji}(t_2) e_c\|),$$

where $\operatorname{dir}(v) = v/\|v\|$ for $0 \neq v \in \mathbf{R}^n$. Set

$$\zeta_1(\pi) = \frac{\partial}{\partial t_1}[\zeta(t_1, 0, 0, 0)(\pi)]_{t_1=0},$$

$$\vdots$$

$$\zeta_{234}(\pi) = \frac{\partial^3}{\partial t_2\, \partial t_3\, \partial t_4}[\zeta(0, t_2, t_3, t_4)(\pi)]_{t_2=t_3=t_4=0},$$

$$\zeta_{1234}(\pi) = \frac{\partial^4}{\partial t_1\, \partial t_2\, \partial t_3\, \partial t_4}[\zeta(t_1, t_2, t_3, t_4)(\pi)]_{t_1=t_2=t_3=t_4=0},$$

for all possible combinations of order one through four.

Therefore

$$d\nu(T_l X_{ij} T_c X_{ab})\psi = \frac{\partial^4}{\partial t_1\, \partial t_2\, \partial t_3\, \partial t_4}[\psi \circ \zeta(t_1, t_2, t_3, t_4)]_{t_1=t_2=t_3=t_4=0}$$
$$= d^4\psi(\zeta_1; \zeta_2; \zeta_3; \zeta_4)$$
$$+ d^3\psi[(\zeta_1; \zeta_2; \zeta_{34}) + (\zeta_1; \zeta_3; \zeta_{24}) + (\zeta_1; \zeta_4; \zeta_{23})$$
$$+ (\zeta_2; \zeta_3; \zeta_{14}) + (\zeta_2; \zeta_4; \zeta_{13}) + (\zeta_3; \zeta_4; \zeta_{12})]$$
$$+ d^2\psi[(\zeta_{12}; \zeta_{34}) + (\zeta_{13}; \zeta_{24}) + (\zeta_{14}; \zeta_{23})$$
$$+ (\zeta_1; \zeta_{234}) + (\zeta_2; \zeta_{134}) + (\zeta_3; \zeta_{124}) + (\zeta_4; \zeta_{123})]$$
$$+ d\psi(\zeta_{1234}),$$

where, denoting by $P = P_{\xi \cap \omega^\perp}$ the orthogonal projection of $\mathbf{R}^n$ onto $\xi \cap \omega^\perp$, and by $v \cdot w$ the ordinary scalar product of $v, w \in \mathbf{R}^n$,

$$\zeta_1(\pi) = (0, -Pe_l/s, -\omega \cdot e_l),$$
$$\zeta_2(\pi) = (-X_{ij}\xi, -X_{ij}\omega, 0),$$
$$\zeta_3(\pi) = (0, -Pe_c/s, -\omega \cdot e_c),$$
$$\zeta_4(\pi) = (-X_{ab}\xi, -X_{ab}\omega, 0),$$
$$\zeta_{12}(\pi) = (0, X_{ij}Pe_l/s, 0),$$
$$\zeta_{13}(\pi) = (0, -[(\omega \cdot e_l)Pe_c + (\omega \cdot e_c)Pe_l + (e_l \cdot Pe_c)\omega]/s^2, e_l \cdot Pe_c/s),$$
$$\zeta_{14}(\pi) = (0, X_{ab}Pe_l/s, 0),$$
$$\zeta_{23}(\pi) = (0, [X_{ij}P - PX_{ij}]e_c/s, -\omega \cdot X_{ij}e_c),$$
$$\zeta_{24}(\pi) = (X_{ab}X_{ij}\xi, X_{ab}X_{ij}\omega, 0),$$
$$\zeta_{34}(\pi) = (0, X_{ab}Pe_c/s, 0),$$

$$\zeta_{123}(\pi) = (0, [(\omega \cdot e_c)X_{ij}Pe_l + (\omega \cdot e_l)[X_{ij}P - PX_{ij}]e_c - (\omega \cdot X_{ij}e_c)Pe_l$$
$$- (e_l \cdot PX_{ij}e_c)\omega + (e_l \cdot Pe_c)X_{ij}\omega]/s^2, e_l \cdot PX_{ij}e_c/s),$$
$$\zeta_{124}(\pi) = (0, -X_{ab}X_{ij}Pe_l/s, 0),$$
$$\zeta_{134}(\pi) = (0, X_{ab}[(\omega \cdot e_l)Pe_c + (\omega \cdot e_c)Pe_l + (e_l \cdot Pe_c)\omega]/s^2, 0),$$
$$\zeta_{234}(\pi) = (0, -X_{ab}[X_{ij}P - PX_{ij}]e_c/s, 0),$$
$$\zeta_{1234}(\pi) = (0, -X_{ab}[(\omega \cdot e_c)X_{ij}Pe_l + (\omega \cdot e_l)[X_{ij}P - PX_{ij}]e_c - (\omega \cdot X_{ij}e_c)Pe_l$$
$$- (e_l \cdot PX_{ij}e_c)\omega + (e_l \cdot Pe_c)X_{ij}\omega]/s^2, 0),$$

where we stipulated that

$$-X_{ij}\xi = \frac{\partial}{\partial t}[r_{ij}(t)\xi]_{t=0},$$

$$X_{ab}X_{ij}\xi = \frac{\partial^2}{\partial t_1 \partial t_2}[r_{ab}(t_2)r_{ij}(t_1)\xi]_{t_1=t_2=0}.$$

We obtain the operator D as

$$D = d\nu(V)$$
$$= \sum_{1 \le i < j < l \le n} [d\nu(T_i X_{jl}T_i X_{jl}) + d\nu(T_i X_{jl}T_j X_{li}) + d\nu(T_i X_{jl}T_l X_{ij})$$
$$+ d\nu(T_j X_{li}T_i X_{jl}) + d\nu(T_j X_{li}T_j X_{li}) + d\nu(T_j X_{li}T_l X_{ij})$$
$$+ d\nu(T_l X_{ij}T_i X_{jl}) + d\nu(T_l X_{ij}T_j X_{li}) + d\nu(T_l X_{ij}T_l X_{ij})].$$

As to D_{ijl} we simply have

$$D_{ijl} = d\nu(V_{ijl}) = d\nu(T_i X_{jl}) + d\nu(T_j X_{li}) + d\nu(T_l X_{ij})$$

for $i, j, l = 1, \ldots, n$, where

$$d\nu(T_l X_{ij})\psi = \frac{\partial^2}{\partial t_1 \partial t_2}[\psi \circ \zeta(t_1, t_2, 0, 0)]_{t_1=t_2=0} = d^2\psi(\zeta_1; \zeta_2) + d\psi(\zeta_{12}),$$

and ζ_1, ζ_2, ζ_{12} are given above.

Pullback to the hyperbolic space. As described in Section 4, the pullback of the operator D is $C = \Psi^{-1}D\Psi$, so its effect on $\phi \in \mathcal{S}(\Gamma)$ will be

$$C\phi = \sigma^{-1}D([\sigma\phi] \circ \tau_k^{-1}) \circ \tau_k.$$

The summands are

$$
\begin{aligned}
\Psi^{-1} & d\nu(T_l X_{ij} T_c X_{ab})\Psi\phi \\
&= d^4\phi(\theta_1; \theta_2; \theta_3; \theta_4) \\
&\quad + d^3\phi[(\theta_1; \theta_2; \theta_{34}) + (\theta_1; \theta_3; \theta_{24}) + (\theta_1; \theta_4; \theta_{23}) \\
&\qquad + (\theta_2; \theta_3; \theta_{14}) + (\theta_2; \theta_4; \theta_{13}) + (\theta_3; \theta_4; \theta_{12})] \\
&\quad + d^2\phi[(\theta_{12}; \theta_{34}) + (\theta_{13}; \theta_{24}) + (\theta_{14}; \theta_{23}) \\
&\qquad + (\theta_1; \theta_{234}) + (\theta_2; \theta_{134}) + (\theta_3; \theta_{124}) + (\theta_4; \theta_{123})] \\
&\quad + d\phi(\theta_{1234}) \\
&\quad + \tanh r\,[d^3\phi[(\theta_2; \theta_3'; \theta_4)\theta_1^r + (\theta_1'; \theta_2; \theta_4)\theta_3^r] \\
&\qquad + d^2\phi[(\theta_2; \theta_{34})\theta_1^r + (\theta_3'; \theta_{24})\theta_1^r + (\theta_1'; \theta_{24})\theta_3^r + (\theta_4; \theta_{23}')\theta_1^r \\
&\qquad\quad + (\theta_1'; \theta_4)\theta_{23}^r + (\theta_2; \theta_{14})\theta_3^r + (\theta_2; \theta_4)\theta_{13}^r + (\theta_4; \theta_{12})\theta_3^r] \\
&\qquad + d\phi[(\theta_{24})\theta_{13}^r + (\theta_{14})\theta_{23}^r + (\theta_{234})\theta_1^r + (\theta_{124})\theta_3^r + (\theta_4)\theta_{123}^r]] \\
&\quad + (1 + 2\tanh^2 r)\,[d^2\phi(\theta_2; \theta_4)\theta_1^r\theta_3^r + d\phi[(\theta_{24})\theta_1^r\theta_3^r + (\theta_4)\theta_1^r\theta_{23}^r]],
\end{aligned}
$$

where θ_J, for each multiindex $J \subset \{1, 2, 3, 4\}$, is obtained from ζ_J by substituting in its expression the variable s with $\tanh r$ and multiplying by $\cosh^2 r$ its third component, which is then denoted as θ_J^r; and θ_J' is obtained from θ_J by doubling its third component. Thus for instance

$$
\begin{aligned}
\theta_1 &= (0, -Pe_l/\tanh r, -\omega \cdot e_l \cosh^2 r), \\
\theta_1' &= (0, -Pe_l/\tanh r, -2\omega \cdot e_l \cosh^2 r), \\
\theta_1^r &= -\omega \cdot e_l \cosh^2 r.
\end{aligned}
$$

Observe that θ_J^r vanishes (and consequently $\theta_J' = \theta_J$) unless

$$
J = \{1\}, \{3\}, \{1, 3\}, \{2, 3\}, \{1, 2, 3\};
$$

the above sum does not appear symmetric in the indices $1, 2, 3, 4$ because the terms that vanish were omitted, and because θ_J has been replaced by θ_J' only in case they were different. Furthermore the factor θ_J^r, when appearing in a summand, is meant to multiply the entire corresponding differential, not its argument: thus

$$
d^3\phi[(\theta_2; \theta_3'; \theta_4)\theta_1^r + (\theta_1'; \theta_2; \theta_4)\theta_3^r]
$$

stands for

$$
\theta_1^r d^3\phi(\theta_2; \theta_3'; \theta_4) + \theta_3^r d^3\phi(\theta_1'; \theta_2; \theta_4).
$$

The whole operator C is now obtained in the same way as D in the preceding section. Analogously we obtain the pullback C_{ijl} of the operator D_{ijl} from

$$
\Psi^{-1} d\nu(T_l X_{ij})\Psi\phi = d^2\psi(\theta_1; \theta_2) + d\psi(\theta_{12}) + \tanh r\,\theta_1^r d\psi(\theta_2).
$$

REFERENCES

[BC1] C. A. Berenstein, E. Casadio Tarabusi, *Inversion formulas for the k-dimensional Radon transform in real hyperbolic spaces*, Duke Math. J. **62** (1991), 613–631.

[BC2] ⸻, *Range of the k-dimensional Radon transform in real hyperbolic spaces* (to appear).

[BZ] C. A. Berenstein, L. Zalcman, *Pompeiu's problem on symmetric spaces*, Comment. Math. Helvetici **55** (1980), 593–621.

[GS] I. M. Gel'fand, G. E. Shilov, *Generalized functions, I: Properties and operations*, Academic Press, New York, 1964.

[G] F. B. Gonzalez, *Invariant differential operators and the range of the Radon d-plane transform*, Math. Ann. **287** (1990), 627–635.

[GR] I. S. Gradshteyn, I. M. Ryzhik, *Table of integrals, series, and products*, corrected and enlarged edition, Academic Press, Orlando, 1980.

[H1] S. Helgason, *Differential operators on homogeneous spaces*, Acta Math. **102** (1959), 239–299.

[H2] ⸻, *Groups and geometric analysis: integral geometry, invariant differential operators, and spherical functions*, Pure and Appl. Math., vol. 113, Academic Press, Orlando, 1984.

[H3] ⸻, *The totally-geodesic Radon transform on constant curvature spaces*, Contemp. Math. **113** (1990), 141–149.

[HOW] J. Horváth, N. Ortner, P. Wagner, *Analytic continuation and convolution of hypersingular higher Hilbert-Riesz kernels*, J. Math. Anal. Appl. **123** (1987), 429–447.

[J] F. John, *The ultrahyperbolic differential equation with 4 independent variables*, Duke Math. J. **4** (1938), 300–322.

[K1] Á. Kurusa, *A characterization of the Radon transform's range by a system of PDEs*, J. Math. Anal. Appl. **161** (1991), 218–226.

[K2] ⸻, *Support theorems for totally geodesic Radon transforms on constant curvature spaces*, Proc. Amer. Math. Soc. (to appear).

[Rr] F. Richter, *Differentialoperatoren auf euklidischen k-Ebenraumen und Radon-transformationen*, Dissertation, Humboldt Universität zu Berlin, 1986.

[Rz] M. Riesz, *L'intégrale de Riemann-Liouville et le problème de Cauchy*, Acta Math. **81** (1949), 1–223.

[W] P. Wagner, *Bernstein-Sato-Polynome und Faltungsgruppen zu Differentialoperatoren*, Z. Anal. Anwendungen **8** (1989), 407–423.

DEPARTMENT OF MATHEMATICS AND SYSTEMS RESEARCH CENTER OF THE UNIVERSITY OF MARYLAND, COLLEGE PARK, MARYLAND 20742, U.S.A.

E-mail address: cab@math.umd.edu, carlos@src.umd.edu

DIPARTIMENTO DI MATEMATICA, UNIVERSITÀ DI TRENTO, 38050 POVO (TRENTO), ITALY

E-mail address: casadio@itnvax.cineca.it, casadio@itncisca.bitnet

Contemporary Mathematics
Volume **140**, 1992

Holmgren's uniqueness theorem and support theorems for real analytic Radon transforms

JAN BOMAN

1. Introduction. If f is a continuous function on $\mathbb{R}^n$ decaying at infinity faster than any negative power of $|x|$, and the integral of f is zero over all hyperplanes not intersecting a given compact convex set K, then f must vanish outside K; this is the well-known support theorem of Helgason [**He1**], [**He2**]. Here we shall study the corresponding problem when f is only assumed to decay fast in a certain open cone Γ in $\mathbb{R}^n$. It turns out that f must then vanish in the set

$$(1) \qquad \bigcap_{x \in K} (x + \Gamma \cup (-\Gamma));$$

see Corollary 3. Examples show that the set (1) is the largest set for which this statement is true. More generally we consider a generalized Radon transform

$$(2) \qquad R_\rho f(H) = \int_H f(x)\rho(x, H)ds,$$

assuming the weight function ρ is positive and real analytic and also real analytic at infinity in a certain sense (see Corollary 2); here ds is the Euclidean surface measure on the hyperplane H, and integration is to be performed with respect to x. It will be convenient to formulate our main theorem in a projective setting; the decay condition then means that f is flat on an open subset of a hyperplane in the projective space $\mathbb{P}^n$.

The proof depends on microlocal regularity properties of solutions to the equation $R_\rho f = 0$ together with a recent vanishing theorem [**B2**] stating that if f is flat along a real analytic hypersurface S and all conormals to S are absent in the analytic wave front set $\mathrm{WF}_A(f)$ of f, then f must vanish in some

1991 *Mathematics Subject Classification.* Primary 44A12; Secondary 35A27.

The author was supported in part by a grant from Swedish Natural Science Research Council

This paper is in final form and no version of it will be submitted for publication elsewhere.

neighborhood of S. This theorem is closely related to a well-known vanishing theorem of Hörmander, which is a crucial part of the elegant proof of Holmgren's uniqueness theorem given in [**Hö1**] (see also [**Hö2**], section 8.6). As in the proof of Holmgren's theorem, in our proof we need to construct a family of "non-characteristic" surfaces, covering the claimed zero-set of f. In our problem the characteristic set consists of all conormals to planes intersecting the set K. Support theorems for generalized Radon transforms have been proved by similar methods in [**BQ1**], [**BQ2**], [**B1**], [**GQ**], [**Q2**], [**Q3**].

2. The support theorem. To describe the zero set of f in the projective setting we shall need to introduce the set $\mathrm{sh}_K(E)$, the shadow of E with respect to K. Denote the n-dimensional real projective space by $\mathbb{P}^n$, and if x and y are two distinct points of $\mathbb{P}^n$, let $G(x,y)$ be the geodesic through x and y with the point x removed. If E is a subset of $\mathbb{P}^n \setminus \{x\}$, we define $P(x,E)$ as the union of all $G(x,y)$ (considered as subsets of $\mathbb{P}^n$) for $y \in E$. If K is a non-empty subset of $\mathbb{P}^n$ and $K \cap E = \emptyset$, we define $\mathrm{sh}_K(E)$ as the intersection

$$\bigcap_{x \in K} P(x, E).$$

Equivalently, $\mathrm{sh}_K(E)$ is equal to the complement in $\mathbb{P}^n \setminus K$ of the union of all geodesics intersecting K but not intersecting E. At least in simple cases $\mathrm{sh}_K(E) \setminus E$ consists of the two shadows of E obtained by letting K act as a light source and assuming light propagates along geodesics in one or the other direction. If E is an open set or an open subset of a hyperplane disjoint from K, then $\mathrm{sh}_K(E)$ is open. Furthermore, it is easy to see that sh_K is a hull operation, that is, $\mathrm{sh}_K(E) = \mathrm{sh}_K(\mathrm{sh}_K(E))$ for any $E \subset \mathbb{P}^n \setminus K$.

Let K and E be disjoint subsets of $\mathbb{R}^n$, K non-empty. Identifying $\mathbb{R}^n$ with a subset of $\mathbb{P}^n$ we can define $\mathrm{sh}_K(E)$ as before. In some cases $\mathrm{sh}_K(E)$ will contain points at infinity. Specifically, if K and E are disjoint compact convex subsets of $\mathbb{R}^n$ with nonempty interior, then the restriction to $\mathbb{R}^n$ of $\mathrm{sh}_K(E) \setminus E$ will have two or three components, one between E and K, one "behind" E as viewed from K, and sometimes one on the opposite side of K from that of E. In the second case the last two components are unbounded; if we move up to projective space they are of course parts of the same component.

Let $\mathcal{G}_n$ denote the set of hyperplanes in $\mathbb{P}^n$. We now choose, for every $H \in \mathcal{G}_n$, a measure $d\sigma$ on H as follows. Using a choice of coordinates $(x_0; x_1; \ldots; x_n)$ on $\mathbb{P}^n$ we obtain a $2 - 1$ map $S^n \to \mathbb{P}^n$ and similarly a $2 - 1$ map $S^{n-1} \to H$ for every $H \in \mathcal{G}_n$. We define $d\sigma$ to be the push-forward of the Euclidean surface measure on S^{n-1} under that map. For a continuous function f on $\mathbb{P}^n$ we can now define our Radon transform R_ρ by

(3) $$R_\rho f(H) = \int_H f(x)\rho(x, H)d\sigma, \quad H \in \mathcal{G}_n.$$

The weight function ρ is defined on the manifold Z consisting of all pairs (x, H) where $x \in H$ and $H \in \mathcal{G}_n$. It is clear that Z is a real analytic submanifold of the product space $\mathbb{P}^n \times \mathcal{G}_n$. We shall say that the function f is *flat* on $E \subset \mathbb{P}^n$ if $f(x)$ tends to zero faster than any power of the distance to E as x tends to E.

If L is a hyperplane in $\mathbb{P}^n$ and K is a subset of $\mathbb{P}^n \setminus L$ we shall say that K is a *convex* subset of $\mathbb{P}^n \setminus L$ if K is the image of a convex subset of $\mathbb{R}^n$ under some affine bijection $\mathbb{R}^n \to \mathbb{P}^n \setminus L$. This property is obviously independent of the choice of the hyperplane L.

THEOREM. *Assume that the weight function ρ is real analytic and positive on Z. Let L be a hyperplane in $\mathbb{P}^n$ and let K be a compact convex subset of $\mathbb{P}^n \setminus L$. Let f be continuous on $\mathbb{P}^n \setminus K$, and assume $R_\rho f(H) = 0$ for all $H \in \mathcal{G}_n$ not intersecting K. Let E be an open subset of L and assume f is flat on E. Then $f = 0$ on $\mathrm{sh}_K(E)$.*

REMARK. It is sufficient to assume that f tends to zero fast as x approaches E from *one side* in some neighborhood of an arbitrary point of E; see remark after Proposition 2.

COROLLARY 1. *With the same hypotheses as in the theorem, assume f is flat on all of L, i.e., $E = L$. Then $f = 0$ in the complement of K.*

PROOF. If E is an entire hyperplane not intersecting K, then $\mathrm{sh}_K(E)$ is equal to the complement of K.

In the remaining corollaries we shall assume given a real analytic Radon transform R_ρ on $\mathbb{R}^n$. Then $\rho(x, H)$ is defined on the manifold Z_0 consisting of all pairs (x, H), $x \in \mathbb{R}^n$, H hyperplane in $\mathbb{R}^n$. An affine bijection $\mathbb{R}^n \to \mathbb{P}^n \setminus L$ for some fixed $L \in \mathcal{G}_n$, for instance the map $\alpha : (x_1, \ldots, x_n) \mapsto (1; x_1; \ldots; x_n)$, induces an identification of Z_0 with a dense subset of our manifold $Z \subset \mathbb{P}^n \times \mathcal{G}_n$. This identification depends of course on the choice of imbedding of $\mathbb{R}^n$ into $\mathbb{P}^n$, but the assumptions in the following corollary do not depend on this choice.

COROLLARY 2. *(Helgason's theorem for real analytic densities, [B1].) Let R_ρ be a real analytic Radon transform in $\mathbb{R}^n$, and assume $\rho(x, H)$ can be extended to a real analytic and positive function on Z. Let K be a compact convex subset of $\mathbb{R}^n$. Let f be continuous on $\mathbb{R}^n \setminus K$ and decaying faster than any negative power of $|x|$ as $|x|$ tends to infinity, and assume $R_\rho f(H) = 0$ for all H not intersecting K. Then $f = 0$ outside K.*

PROOF. The imbedding α introduced above transforms the measure ds on $H \subset \mathbb{R}^n$ into a measure $\alpha_*(ds) = b(x, H)d\sigma$ on $H \subset \mathbb{P}^n$, where the density $b(x, H)$ factors into a product of one function depending only on x and one depending only on H, $b(x, H) = b_0(x)b_1(H)$ (see Lemma 1 in [B1]). Using this fact we showed in [B1] that the assumptions in Corollary 2 allow us to transform the problem to one in $\mathbb{P}^n$, where f is flat on an entire hyperplane in $\mathbb{P}^n$ (for details we refer to [B1]). The conclusion now follows from Corollary 1.

COROLLARY 3. *Let R_ρ and K be as in Corollary 2. Let Γ be an open conic subset of $\mathbb{R}^n$, let f be continuous on $\mathbb{R}^n \setminus K$ and decaying enough at infinity to be integrable on hyperplanes, for instance $f(x) = \mathcal{O}(|x|^{-n})$ as $|x| \to \infty$. Assume moreover that f decays faster than any negative power of $|x|$ as x tends to infinity in Γ. Assume finally $R_\rho f(H) = 0$ for all H not intersecting K. Then $f = 0$ in the set (1).*

PROOF. Imbedding $\mathbb{R}^n$ in $\mathbb{P}^n$ as in Corollary 2 we obtain the situation in the theorem, or in the remark after it, with $L = H_\infty$, the hyperplane at infinity, and $E = H_\infty \cap \bar{\Gamma}$, where $\bar{\Gamma}$ is the closure of Γ in $\mathbb{P}^n$. In this case $\text{sh}_K(E)$ is the set (1).

COROLLARY 4. *Let R_ρ and K be as in Corollary 2. Let f be continuous and decaying enough at infinity to be integrable on hyperplanes, for instance $f(x) = \mathcal{O}(|x|^{-n})$ as $|x| \to \infty$, and assume $R_\rho f(H) = 0$ for all H not intersecting K. Let L be a hyperplane in $\mathbb{R}^n$ not intersecting K, let E be an open subset of L, and assume f is flat on E. Then $f = 0$ in $\text{sh}_K(E)$.*

PROOF. Imbedding $\mathbb{R}^n$ in $\mathbb{P}^n$ as in Corollary 2 we obtain the statement immediately from the theorem.

REMARK. The function f in the theorem and the corollaries may be any distribution on $\mathbb{P}^n$, provided the flatness condition used here is replaced by an appropriate flatness condition valid for distributions (cf. Proposition 2).

3. The microlocal regularity theorem. Our Radon transform can be written

$$R_\rho f(H) = \int K(x, H) f(x) \, dx,$$

where the distribution $K(x, H)$ on $\mathbb{P}^n \times \mathcal{G}_n$ is a measure supported on the hypersurface Z, in fact a smooth positive density on that surface. The wave front set of K is therefore contained in the conormal manifold $N^*(Z) \subset T^*(\mathbb{P}^n \times \mathcal{G}_n)$ to Z. The corresponding operator is an especially simple kind of Fourier integral operator. Using the natural identification $T^*(\mathbb{P}^n \times \mathcal{G}_n) \simeq T^*(\mathbb{P}^n) \times T^*(\mathcal{G}_n)$ we can consider the wave front set of K, $\text{WF}(K)$, as a subset of $T^*(\mathbb{P}^n) \times T^*(\mathcal{G}_n)$. It is a trivial fact that (cf. [**Hö2**], ch. 8.2)

(4) $\text{WF}(R_\rho f) \subset \{(H, \eta); (x, \xi) \in \text{WF}(f), (x, \xi; H, -\eta) \in \text{WF}(K)\}.$

Using the facts that $\rho > 0$ and that Z has a certain geometric property ($N^*(Z)$ is the graph of an injective map $T^*(\mathbb{P}^n) \mapsto T^*(\mathcal{G}_n)$; see [**GS**]), one can also prove the opposite inclusion; in other words we have equality in (4). Since the manifold Z and the function ρ are real analytic these facts also hold with WF replaced by WF_A (see [**B1**], section 4 and references given there). The manifold $N^*(Z)$ consists of the set of pairs $(x, \xi; H, \eta)$ such that $x \in H$, ξ is conormal to H, η is conormal at H to the hypersurface $\gamma_x = \{L; x \in L\}$ in $\mathcal{G}_n$, and ξ, η are coupled by a certain linear relation; to see this one can locally represent Z as $F(x, H) = 0$ for some smooth function F with non-vanishing gradient and note that $N^*(Z)$ then becomes $\{(x, \lambda F'_x, H, \lambda F'_H);\ F(x, H) = 0, \lambda \in \mathbb{R}\}$. These facts imply the following regularity theorem.

PROPOSITION 1. *Let f be continuous on $\mathbb{P}^n$, or, more generally, let $f \in \mathcal{D}'(\mathbb{P}^n)$, the space of distributions on $\mathbb{P}^n$. Assume $\rho(x, H)$ is real analytic and positive on Z and that $R_\rho f(H) = 0$ for all H in some neighbourhood of $H_0 \in \mathcal{G}_n$. Then*

$$N^*(H_0) \cap \mathrm{WF_A}(f) = \emptyset.$$

4. Vanishing theorems for microanalytic distributions. Assume for a moment that we knew about our function f that $\mathrm{WF_A}(f) \cap N^*(S) = \emptyset$ for some *closed* real analytic surface S, for instance a sphere in $\mathbb{R}^n$ with radius 1, and that f is flat along S. Then it would follow almost immediately from the definition of the analytic wave front set that the one-variable function $u(t)$ defined as the integral of f over a family of concentric spheres S_t with radius t must be real analytic and flat at $t = 1$, hence vanish identically near that point. Since the same would be true with f replaced by fh for an arbitrary real analytic function h, it would follow that $f = 0$ in a neighborhood of S. In [**B1**], where we assumed f tends to zero fast in all directions at infinity, we had that situation. In this paper, however, our function is only assumed to be flat along some open piece of a hyperplane in $\mathbb{P}^n$; it is clear that the argument just outlined does not apply then. However, the vanishing theorem that we need is actually true [**B2**]. Note that the restriction to a smooth submanifold $S \subset \mathbb{R}^n$ of a distribution f defined in some neighborhood of S is well defined if $N^*(S) \cap \mathrm{WF}(f)$ is empty.

PROPOSITION 2, [**B2**]. *Let f be a distribution defined in some neighborhood of the real analytic hypersurface $S \subset \mathbb{R}^n$, and assume*

(5) $$N^*(S) \cap \mathrm{WF_A}(f) = \emptyset.$$

Assume the restrictions to S of f and all its derivatives vanish, or if f is continuous, assume equivalently that $f(x)$ tends to zero faster than any power of the distance from x to S as x tends to S. Then $f = 0$ in some neighbourhood of S.

REMARKS. If f is continuous, it is sufficient to assume $f(x)$ tends to zero fast as x approaches S from one side. If f is assumed to *vanish* in a one-sided neighborhood of the surface S, it is sufficient to assume S is C^1 and piecewise C^2, because then we can find a ball inside the zero-set of f, tangent to S at an arbitrary point of S; this observation will be useful in the proof below. In this case it is also sufficient, by Hörmander's vanishing theorem referred to above, to assume *one* of the two conormals $(x, \pm\xi)$ absent in $\mathrm{WF_A}(f)$. For related and sharper theorems, first proved by Kashiwara, see [**Hö2**], ch. 9.6. Those theorems are valid for hyperfunctions; however, the assertion of Proposition 2 is *not* valid for hyperfunctions (see [**K**], note 3.3).

5. Proof of the theorem. Let f satisfy the assumptions of the theorem. Since the hyperplane L is disjoint from K, we infer immediately from Proposition

1 that $N^*(L) \cap \mathrm{WF}_A(f)$ is empty. Working locally near an arbitrary point of E we can then conclude from Proposition 2 that f must vanish in some neighborhood of E in $\mathbb{P}^n$. To prove that f must vanish on $\mathrm{sh}_K(E)$ we are going to reduce our problem to an especially simple situation in $\mathbb{R}^n$ by choosing a convenient hyperplane $M \subset \mathbb{P}^n$ and using $\mathbb{R}^n \cup H_\infty$ as a model for $\mathbb{P}^n$ with $M = H_\infty$. Let z be an arbitrary point of $\mathrm{sh}_K(E) \setminus L$. Since K is a convex subset of $\mathbb{P}^n \setminus L$ and $z \notin K$, we can find a hyperplane H through z that does not meet K. Let $\mathcal{J}$ be the set of hyperplanes containing $H \cap L$, that are different from L and do not intersect K. Then $\mathcal{J}$ is a one-dimensional manifold with two components; one of those components contains no plane through z. Choose the hyperplane M in that component and represent $\mathbb{P}^n$ as indicated above. Then K is a compact convex subset of $\mathbb{R}^n \subset \mathbb{R}^n \cup H_\infty$. Choose a linear form ϕ on $\mathbb{R}^n$ which vanishes on $L \cap \mathbb{R}^n$ and is positive on K. The planes in $\mathcal{J}$ now take the form $\phi(x) = c$, and our choice of M is easily seen to imply that z lies between L and K, i.e., that $0 < \phi(z) < \inf_{x \in K} \phi(x)$. Hence the set $F_z = L \cap P(z, K)$ must be a bounded subset of $L \cap \mathbb{R}^n$. We shall prove that $f = 0$ near z by constructing a family of surfaces filling a region between E and z and whose tangent planes are disjoint from K.

From now on all our reasoning will take place in $\mathbb{R}^n$ and we shall denote $L \cap \mathbb{R}^n$ by L. Since F_z is a convex compact subset of E we can take a bounded convex set V, open in L, containing F_z, with closure $\bar{V}$ contained in E, and with smooth boundary $\partial_L(V)$ as a subset of L. Since $z \in \mathrm{sh}_K(E)$, any line through z and $\partial_L(V)$ must be disjoint from K, and since V is convex, any hyperplane through z, tangent to $\partial_L(V)$, must be disjoint from K. Choose a ball $B_\varepsilon(z)$ with center at z and radius ε so small that all common tangent planes to $\partial_L V$ and $B_\varepsilon(z)$ are also disjoint from K, and ε is smaller than the distance from z to L. Let $z_0 \in V$, set $z_t = z_0 + t(z - z_0)$ for $0 < t \leq 1$, and let Σ_t be the boundary of the convex hull of $B_{\varepsilon t}(z_t) \cup V$ with all points on L removed. Then each Σ_t is piecewise smooth, and using the fact that $B_{\varepsilon t}(z_t)$ is contained in the convex hull of $B_\varepsilon(z) \cup V$ it is easy to see that Σ_t is non-characteristic in the sense that all tangent planes to Σ_t are disjoint from K.

Assume now that $z \in \mathrm{supp} f$. Since $\bar{V} \subset E$, there must be a $\bar{t}$, $0 < \bar{t} \leq 1$, such that Σ_t is disjoint from the support of f for $t < \bar{t}$ and $\Sigma_{\bar{t}}$ meets the support of f. By Proposition 1 and the remark following Proposition 2 this is impossible; hence we have obtained a contradiction and can conclude that $z \notin \mathrm{supp} f$. Since z was an arbitrary point of $\mathrm{sh}_K(E)$ we have proved that $f = 0$ in $\mathrm{sh}_K(E)$.

6. Counterexamples. To show that the decay assumption in Helgason's original theorem cannot be omitted Helgason gave the examples $\Re(x_1 + ix_2)^{-m}$, $m = 2, 3, \ldots$. More generally, Quinto characterized the null space of the Radon transform on (weighted) L^2-space on the complement of a ball in $\mathbb{R}^n$, [**Q1**]. Helgason's examples can also be generalized as follows. Let $g(x)$ be a distribution in $\mathbb{R}^n$ which is positively homogeneous of degree $-n - k$ (as a distribution in $\mathbb{R}^n$), k integer ≥ 0, and C^∞ in $\mathbb{R}^n \setminus \{0\}$. Furthermore we assume g is even if

k is even and odd if k is odd. Write $Rf(\omega, p) = Rf(H_{(\omega,p)})$, where $H_{(\omega,p)}$ is the hyperplane $\{x; x \cdot \omega = p\}$, $(\omega, p) \in S^{n-1} \times \mathbb{R}$. The n-dimensional Fourier transform $\hat{g}$ of g is C^∞ outside the origin and homogeneous of degree k, hence a locally bounded function. The one-dimensional Fourier transform of $Rg(\omega, p)$ with respect to p, which we denote by $\widehat{Rg}(\omega, \tau)$, is connected with g by the familiar formula $\widehat{Rg}(\omega, \tau) = \hat{g}(\tau\omega)$. Since $\hat{g}$ is even if k is even and odd if k is odd, we then obtain

$$\widehat{Rg}(\omega, \tau) = \hat{g}(\tau\omega) = \tau^k \hat{g}(\omega), \quad \tau \in \mathbb{R},$$

and (we denote the Dirac measure at the origin by δ)

$$Rg(\omega, p) = i^k \delta^{(k)}(p)\hat{g}(\omega),$$

which means in particular that $Rg(H) = 0$ for all hyperplanes H not containing the origin. Thus, to generalize Helgason's example we can take any non-trivial smooth function g in $\mathbb{R}^n \setminus \{0\}$ which has the correct parity, is homogeneous of degree $-n - k$ and *can be extended to a homogeneous distribution in $\mathbb{R}^n$*. It is important to note that such extension is not always possible (see [**Hö2**], ch. 3). For instance, if $k = 0$ the necessary and sufficient condition is that the integral of g over a sphere centered at the origin is zero. Since the product of a homogeneous distribution and a homogeneous polynomial is homogeneous, it follows that for $k \geq 1$ the product pg, for any homogeneous polynomial of degree k, must have integral zero over such spheres. This condition is also sufficient. In other words, for $k = 0$ we need to take g homogeneous of degree $-n$, even, and with mean zero.

To prove that the set (1) in Corollary 3 cannot be replaced by any larger set we take g homogeneous of degree $-n$, even, with mean zero, equal to zero in the symmetric cone Γ, and choose

$$f(x) = \int_K g(x - y)dy, \quad x \notin K.$$

Then f vanishes in the set $\cap_{x \in K}(x + \Gamma)$, and in general f does not vanish in any larger set. If the set K has interior points we can of course make f continuous in $\mathbb{R}^n$ by taking a non-trivial smooth function ψ, supported in K, and choosing $f(x) = \text{p.v.} \int_{\mathbb{R}^n} g(x - y)\psi(y)dy$, $x \in \mathbb{R}^n$; here "p.v." indicates that we must take the principal value of the divergent integral. By transferring these examples to $\mathbb{P}^n$ we see that the set $\text{sh}_K(E)$ cannot be replaced by any larger set in the conclusion of the theorem.

REFERENCES

[B1]   J. Boman, *Helgason's support theorem for Radon transforms — a new proof and a generalization*, Mathematical methods in Tomography, Proceedings of a conference held at Oberwolfach 1990, Springer Lecture notes in Mathematics no. 1497, 1991, pp. 1-5.

[B2]   J. Boman, *A local vanishing theorem for distributions*, C. R. Acad. Sci. Paris (to appear).

[BQ1] J. Boman and E. T. Quinto, *Support theorems for real-analytic Radon transforms*, Duke Math. J. **55** (1987), 943-948.

[BQ2] J. Boman and E. T. Quinto, *Support theorems for real-analytic Radon transforms on line complexes in three-space*, Trans. Amer. Math. Soc. (to appear).

[GQ]   F. Gonzalez and E. T. Quinto, *Support theorems for Radon transforms on higher rank symmetric spaces*, preprint (1991).

[GS]   V. Guillemin and S. Sternberg, *Geometric Asymptotics*, Amer. Math. Soc., Providence, RI, 1977.

[He1] S. Helgason, *The Radon transform on Euclidean spaces, compact two-point homogeneous spaces and Grassmann manifolds*, Acta Math. **113** (1965), 153-180.

[He2] S. Helgason, *The Radon transform*, Birkhäuser, Boston, 1980.

[Hö1] L. Hörmander, *Uniqueness theorems and wave front sets for solutions of linear differential equations with analytic coefficients*, Comm. Pure Appl. Math. **24** (1971), 671-704.

[Hö2] L. Hörmander, *The analysis of linear partial differential operators, vol. 1*, Springer-Verlag, Berlin, Heidelberg, and New York, 1983.

[K]   A. Kaneko, *Introduction to hyperfunctions*, Kluwer Academic Publishers, Dordrecht, Boston, London, 1988.

[Q1]   E. T. Quinto, *Null spaces and ranges for classical and spherical Radon transforms*, J. Math. Anal. Appl. **90** (1982), 408-420.

[Q2]   E. T. Quinto, *Real analytic Radon transforms on rank one symmetric spaces*, Proc. Amer. Math. Soc. (1991) (to appear).

[Q3]   E. T. Quinto, *A note on flat Radon transforms*, in this volume.

DEPARTMENT OF MATHEMATICS, STOCKHOLM UNIVERSITY, BOX 6701, S-11385 STOCKHOLM, SWEDEN

E-mail address: jabo@matematik.su.se

Contemporary Mathematics
Volume **140**, 1992

On the Petty–Schneider Theorem

G.D. CHAKERIAN AND E. LUTWAK

A convex figure in Euclidean n–space, $\mathbf{R}^n$, is a compact convex subset of $\mathbf{R}^n$. Let $\mathcal{C}^n$ denote the set of convex figures in $\mathbf{R}^n$, and let $\mathcal{C}_s^n$ denote the set of convex figures which have a center of symmetry. A convex body is a convex figure with non–empty interior, and $\mathcal{K}^n$, and $\mathcal{K}_s^n$, will be used to denote the set of convex bodies in $\mathcal{C}^n$ and $\mathcal{C}_s^n$, respectively.

Write $\mathrm{vol}_n(K)$ for the n–dimensional volume of the figure $K \in \mathcal{C}^n$. For a direction $u \in S^{n-1}$, let $u^\perp$ denote the $(n-1)$–dimensional subspace of $\mathbf{R}^n$ that is orthogonal to u. The image of the (orthogonal) projection of a figure $K \in \mathcal{C}^n$ onto $u^\perp$ will be denoted by $K|u^\perp$.

A problem of Shephard [**1964**] asks: If $K, L \in \mathcal{C}_s^n$, and if

$$\mathrm{vol}_{n-1}(K|u^\perp) \leq \mathrm{vol}_{n-1}(L|u^\perp), \qquad \text{for all } u \in S^{n-1},$$

does it follow that $\mathrm{vol}_n(K) \leq \mathrm{vol}_n(L)$?

For $n = 2$, Shephard's question has an easily obtained affirmative answer. Petty [**1967**] and Schneider [**1967**], independently proved that the answer to Shephard's question is negative for all $n > 2$. In a positive direction, Petty and Schneider showed that if attention is restricted to convex figures which have a kind of 'supersymmetry', then the Shephard question has an affirmative answer.

To define this supersymmetry, a bit of additional notation is helpful. Recall that for $K, L \in \mathcal{C}^n$, and $\lambda, \mu \geq 0$ (not both zero), the Minkowski linear combination $\lambda K + \mu L \in \mathcal{C}^n$ is defined by

$$\lambda K + \mu L = \{\, \lambda x + \mu y : x \in K \text{ and } y \in L \,\}.$$

The Blaschke–Hausdorff distance, $\delta(K, L)$, of $K, L \in \mathcal{C}^n$, can be defined by:

$$\delta(K, L) = \min\{\, \varepsilon \geq 0 : K \subset L + \varepsilon B \text{ and } L \subset K + \varepsilon B \,\},$$

1991 *Mathematics Subject Classification.* 52A40.

Research supported, in part, by NSF Grant DMS–8902550

This paper is in final form and no version of it will be submitted for publication elsewhere.

where B denotes the unit ball in $\mathbf{R}^n$. The class $\mathcal{C}^n$ will be viewed as endowed with the topology induced by the Blaschke–Hausdorff metric. A body in $\mathcal{C}^n$ which is the limit of a Minkowski sum of line segments is called a zonoid, and $\mathcal{Z}^n$ will denote the set of zonoids in $\mathcal{C}^n$. Zonoids are highly symmetric figures. For example, in order for a polytope in $\mathbf{R}^3$ to be a zonoid it must not only be centrally symmetric, but all of its faces must be centrally symmetric as well. See Schneider & Weil [1983] for an excellent survey on zonoids.

The Petty–Schneider theorem (established, independently, by Petty [1967] and Schneider [1967]) is:

THEOREM 1. *If $K \in \mathcal{K}^n$ and $L \in \mathcal{Z}^n$, and if $\mathrm{vol}_{n-1}(K|u^\perp) \leq \mathrm{vol}_{n-1}(L|u^\perp)$, for all $u \in S^{n-1}$, then $\mathrm{vol}_n(K) \leq \mathrm{vol}_n(L)$, with equality if and only if K is a translate of L.*

Schneider [1967] also proved that for each $K \in \mathcal{K}_s^n \backslash \mathcal{Z}^n$, which is sufficiently smooth and has positive curvature, there exists an $L \in \mathcal{K}_s^n \backslash \mathcal{Z}^n$, such that $\mathrm{vol}_{n-1}(K|u^\perp) \leq \mathrm{vol}_{n-1}(L|u^\perp)$, for all $u \in S^{n-1}$, and $\mathrm{vol}_n(K) > \mathrm{vol}_n(L)$. (See Schneider [1967] for details.)

Write $S(Q)$ for the surface area of the figure $Q \in \mathcal{C}^n$. The Cauchy surface area formula gives the surface area, of $Q \in \mathcal{C}^n$, as

$$S(Q) = \frac{1}{\omega_{n-1}} \int_{S^{n-1}} \mathrm{vol}_{n-1}(Q|u^\perp) \, du,$$

where the integration is with respect to Lebesgue measure on S^{n-1}, and ω_i denotes the i–dimensional volume of the unit ball in $\mathbf{R}^i$. Fáry & Makai (see Croft, Falconer & Guy [1991, p. 22]) pose a question similar to Shephard's: If $K, L \in \mathcal{C}_s^n$, and if

$$S(\phi K) \leq S(\phi L), \qquad \text{for all } \phi \in GL(n),$$

does it follow that $\mathrm{vol}_n(K) \leq \mathrm{vol}_n(L)$?

The aim of this note is to show that the question of Fáry and Makai is just another version of Shephard's question. Specifically, the following will be shown:

PROPOSITION 1. *If $K, L \in \mathcal{C}^n$ then*

$$\mathrm{vol}_{n-1}(K|u^\perp) \leq \mathrm{vol}_{n-1}(L|u^\perp), \qquad \text{for all } u \in S^{n-1},$$

if and only if

$$S(\phi K) \leq S(\phi L), \qquad \text{for all } \phi \in GL(n).$$

This provides the following equivalent formulation of the Petty–Schneider theorem:

THEOREM 2. *If $K \in \mathcal{K}^n$ and $L \in \mathcal{Z}^n$, and if $S(\phi K) \leq S(\phi L)$, for all $\phi \in GL(n)$, then $\mathrm{vol}_n(K) \leq \mathrm{vol}_n(L)$, with equality if and only if K is a translate of L.*

To prove the proposition, some well–known results regarding mixed volumes will be used. The mixed volume, $V_1(K, L)$, of $K, L \in \mathcal{C}^n$, may be defined by:

$$nV_1(K, L) = \lim_{\varepsilon \to 0^+} \frac{\mathrm{vol}_n(K + \varepsilon L) - \mathrm{vol}_n(K)}{\varepsilon}.$$

Recall that for $K \in \mathcal{C}^n$, the surface area $S(K) = nV_1(K, B)$, where B denotes the unit ball in $\mathbf{R}^n$. The following four properties of the mixed volume V_1 will be used. (Proofs can be found in Bonnesen & Fenchel [**1934**] or Leichtweiß [**1980**].)

For $Q \in \mathcal{C}^n$, the mixed volume, $V_1(Q, \cdot)$, is Minkowski linear; i.e.,

$$V_1(Q, \lambda K + \mu L) = \lambda V_1(Q, K) + \mu V_1(Q, L). \tag{1}$$

The mixed volume V_1 is invariant under independent translations of the bodies, and simultaneous $SL(n)$–transformations of the bodies; i.e., for $K, L \in \mathcal{C}^n$, and $x, y \in \mathbf{R}^n$,

$$V_1(x + \phi K, y + \phi L) = V_1(K, L), \qquad \text{for all } \phi \in SL(n). \tag{2}$$

Also used will be the fact that for $Q \in \mathcal{C}^n$,

$$V_1(Q, \cdot) \colon \mathcal{C}^n \longrightarrow \mathbf{R}, \qquad \text{is continuous.} \tag{3}$$

The last fact needed is that for $K \in \mathcal{C}^n$ and $u \in S^{n-1}$,

$$nV_1(K, \bar{u}) = \mathrm{vol}_{n-1}(K|u^{\perp}), \tag{4}$$

where $\bar{u}$ denotes the line segment connecting the origin and the point u.

Now, suppose that $K, L \in \mathcal{C}^n$ are such that $\mathrm{vol}_{n-1}(K|u^{\perp}) \leq \mathrm{vol}_{n-1}(L|u^{\perp})$, for all $u \in S^{n-1}$. Since zonoids are limits of sums of line segments, it follows from this and properties (4), (1), (3), and (2), that

$$V_1(K, Z) \leq V_1(L, Z), \qquad \text{for all } Z \in \mathcal{Z}^n.$$

Since all ellipsoids are zonoids, it follows that

$$V_1(K, \phi^{-1}B) \leq V_1(L, \phi^{-1}B), \qquad \text{for all } \phi \in SL(n).$$

From this, the fact that $S(Q) = nV_1(Q, B)$, and properties (1) and (2), it follows that,

$$S(\phi K) \leq S(\phi L), \qquad \text{for all } \phi \in SL(n),$$

and consequently for all $\phi \in GL(n)$.

On the other hand, suppose that $K, L \in \mathcal{C}^n$ are such that

$$S(\phi K) \leq S(\phi L), \qquad \text{for all } \phi \in GL(n).$$

From the fact that $S(Q) = nV_1(Q, B)$, and properties (1) and (2), it follows that

$$V_1(K, E) \leq V_1(L, E), \qquad \text{for all ellipsoids } E.$$

Given $u \in S^{n-1}$, choose a sequence of nondegenerate ellipsoids converging to the segment $\bar{u}$, and use property (3) to conclude:

$$V_1(K, \bar{u}) \leq V_1(L, \bar{u}), \qquad \text{for all } u \in S^{n-1}.$$

Property (4) now gives:

$$\text{vol}_{n-1}(K|u^{\perp}) \leq \text{vol}_{n-1}(L|u^{\perp}), \qquad \text{for all } u \in S^{n-1},$$

and completes the proof of the proposition.

A natural extension of the proposition can be given. For $Q \in C^n$, the Quermassintegrals $W_0(Q), W_1(Q), \ldots, W_n(Q)$ of Q, can be defined by letting, $W_0(Q) = \text{vol}_n(Q)$, $W_n(Q) = \omega_n$, and for $0 < i < n$

$$W_{n-i}(Q) = \frac{\omega_n}{\omega_i} \int_{G(n,i)} \text{vol}_i(Q|\xi) \, d\xi,$$

where the integration is over $G(n,i)$, the Grassmannian of i-dimensional subspaces of $\mathbf{R}^n$ with respect to the usual rotation invariant probability measure on $G(n,i)$, and $Q|\xi$ is the image of the orthogonal projection of Q onto $\xi \in G(n,i)$. Thus, $n\,W_1(Q) = S(Q)$, and $(2/\omega_n)W_{n-1}(Q)$, is the mean width of Q.

The following extension of Proposition 1 will be established:

PROPOSITION 1*. *If* $K, L \in C^n$, *and* $0 < i < n$ *then*

$$\text{vol}_i(K|\xi) \leq \text{vol}_i(L|\xi), \qquad \text{for all } \xi \in G(n,i),$$

if and only if

$$W_{n-i}(\phi K) \leq W_{n-i}(\phi L), \qquad \text{for all } \phi \in GL(n).$$

To prove this, mixed volumes more general than V_1 will be used. These mixed volumes are the coefficients of the expansion of the volume of Minkowski linear combinations. Specifically, if $K_1, \ldots, K_r \in C^n$ and $\lambda_1, \ldots, \lambda_r \geq 0$, then the volume of the Minkowski combination $\lambda_1 K_1 + \cdots + \lambda_r K_r$, can be expressed as a symmetric homogeneous n-th degree polynomial in the λ_i:

$$V(\lambda_1 K_1 + \cdots + \lambda_r K_r) = \sum \lambda_{i_1} \cdots \lambda_{i_n} V_{i_1 \cdots i_n}, \tag{5}$$

where the sum is taken over all n-tuples $(i_1, \ldots, i_n)$ of positive integers not exceeding r. The coefficient $V_{i_1 \cdots i_n}$ (which is required to be symmetric in its subscripts) depends only on the figures $K_{i_1}, \ldots, K_{i_n}$, is uniquely determined by (5). This coefficient is called the mixed volume of $K_{i_1}, \ldots, K_{i_n}$, and is written as $V(K_{i_1}, \ldots, K_{i_n})$.

It is well known that the Quermassintegrals of a body $Q \in C^n$ are just special mixed volumes involving Q and the unit ball B. Specifically, for $0 \leq i \leq n$,

$$W_i(Q) = V(\underbrace{Q, \ldots, Q}_{n-i}, \underbrace{B, \ldots, B}_{i}). \tag{6}$$

If $K, L \in \mathcal{C}^n$, then

$$V_1(K, L) = V(\underbrace{K, \ldots, K}_{n-1}, L).$$

The mixed volume

$$V: \underbrace{\mathcal{C}^n \times \cdots \times \mathcal{C}^n}_{n} \longrightarrow [0, \infty),$$

is invariant under independent translations of the bodies and simultaneous unimodular affine transformations of the bodies. It is continuous and Minkowski linear in each of its arguments.

To state the extension of property (4), additional notation is needed: If $u_1, \ldots, u_r \in S^{n-1}$, then write $[u_1, \ldots, u_r]$ for the linear space spanned by $u_1, \ldots, u_r$, and for the orthogonal complement of $[u_1, \ldots, u_r]$ write $[u_1, \ldots, u_r]^\perp$. The extension of property (4) is: If $K \in \mathcal{C}^n$ and $u_1, \ldots, u_r \in S^{n-1}$, then

$$V(\underbrace{K, \ldots, K}_{n-r}, \bar{u}_1, \ldots, \bar{u}_r) = c_r \mathrm{vol}_{n-r}(K|[u_1, \ldots, u_r]^\perp)\mathrm{vol}_r(\bar{u}_1 + \cdots + \bar{u}_r), \quad (7)$$

where c_r depends only on n and r. Property (4) may be extended in another manner: If $\xi \in G(n, r)$, then

$$V(\underbrace{K, \ldots, K}_{n-r}, \underbrace{B \cap \xi, \ldots, B \cap \xi}_{r}) = c'_r \mathrm{vol}_{n-r}(K|\xi^\perp), \quad (8)$$

where c'_r again depends only on n and r. For these results, see Burago–Zalgaller [1988].

To prove Proposition 1*, it will be shown that if $K, L \in \mathcal{C}^n$, and $0 < r < n$, then the following statements are equivalent:

$$W_r(\phi K) \leq W_r(\phi L) \qquad \text{for all } \phi \in GL(n). \quad (9.1)$$

$$\mathrm{vol}_{n-r}(K|\xi) \leq \mathrm{vol}_{n-r}(L|\xi) \qquad \text{for all } \xi \in G(n, n-r). \quad (9.2)$$

$$V(\underbrace{K, \ldots, K}_{n-r}, Z_1, \ldots, Z_r) \leq V(\underbrace{L, \ldots, L}_{n-r}, Z_1, \ldots, Z_r) \qquad \text{for all } Z_i \in \mathcal{Z}^n. \quad (9.3)$$

Suppose (9.1) holds, and $\xi \in G(n, r)$. From (9.1), (6) and the invariance of mixed volumes under simultaneous $SL(n)$–transformations of the bodies (and the Minkowski linearity), it follows that

$$V(\underbrace{K, \ldots, K}_{n-r}, \underbrace{E, \ldots, E}_{r}) \leq V(\underbrace{L, \ldots, L}_{n-r}, \underbrace{E, \ldots, E}_{r}),$$

for any n–dimensional ellipsoid E. Now take a sequence of nondegenerate ellipsoids converging to $B \cap \xi^\perp$, and use the continuity of mixed volumes, together with (8), to get (9.2).

Suppose (9.2) holds. First observe that if $u_1, \ldots, u_r \in S^{n-1}$, then

$$V(\underbrace{K, \ldots, K}_{n-r}, \bar{u}_1, \ldots, \bar{u}_r) \leq V(\underbrace{L, \ldots, L}_{n-r}, \bar{u}_1, \ldots, \bar{u}_r). \quad (10)$$

To see this note that it may be assumed that $u_1, \ldots, u_r \in S^{n-1}$ are linearly independent (if they are not, both sides of (10) are 0). The inequality (10) follows from (9.2) by letting $\xi = [u_1, \ldots, u_r]^\perp$ and using (7). The Minkowski linearity of mixed volumes shows that (10) implies

$$V(\underbrace{K, \ldots, K}_{n-r}, Z_1, \ldots, Z_r) \leq V(\underbrace{L, \ldots, L}_{n-r}, Z_1, \ldots, Z_r),$$

whenever the Z_i can be written as Minkowski sums of segments. The continuity of mixed volumes now gives (9.3) for arbitrary $Z_i \in \mathcal{Z}^n$.

Finally, suppose (9.3) holds and $\phi \in GL(n)$. Taking $Z_1 = \cdots = Z_r = \phi^{-1}B$ in (9.3) results in

$$V(\underbrace{K, \ldots, K}_{n-r}, \underbrace{\phi^{-1}B, \ldots, \phi^{-1}B}_{r}) \leq V(\underbrace{L, \ldots, L}_{n-r}, \underbrace{\phi^{-1}B, \ldots, \phi^{-1}B}_{r}).$$

The invariance of mixed volumes under simultaneous $SL(n)$–transformations of the bodies (and the Minkowski linearity), now gives (9.1).

Schneider [1967] proved a general version of Theorem 1. Specifically, he established:

THEOREM 1*. *If $K \in \mathcal{K}^n$ and $L \in \mathcal{Z}^n$, and $0 < r < n$, and if*

$$\mathrm{vol}_r(K|\xi) \leq \mathrm{vol}_r(L|\xi), \qquad \textit{for all } \xi \in G(n,r),$$

then for all j, such that $0 \leq j < n - r$,

$$W_j(K) \leq W_j(L),$$

with equality for any j, if and only if K and L are translates.

In light of Proposition 1*, Schneider's theorem can be restated as the following extension of Theorem 2:

THEOREM 2*. *If $K \in \mathcal{K}^n$ and $L \in \mathcal{Z}^n$, and $0 < r < n$, and if*

$$W_r(\phi K) \leq W_r(\phi L) \qquad \textit{for all } \phi \in GL(n),$$

then for all j, such that $0 \leq j < r$,

$$W_j(K) \leq W_j(L),$$

with equality for any j, if and only if K is a translate of L.

For the case where $r = n - 1$, the requirement that $L \in \mathcal{Z}^n$, can be relaxed considerably. Specifically, if $K \in \mathcal{K}^n$ and $L \in \mathcal{K}_s^n$, and if

$$W_{n-1}(\phi K) \leq W_{n-1}(\phi L) \qquad \textit{for all } \phi \in GL(n),$$

then for all j, such that $0 \leq j < n - 1$,

$$W_j(K) \leq W_j(L), \tag{11}$$

with equality for any j if and only if K and L are translates.

To see this, note that from Proposition 2*

$$\mathrm{vol}_1(K|\xi) \leq \mathrm{vol}_1(L|\xi), \qquad \text{for all } \xi \in G(n,1),$$

But this implies that there exists an x (the center of symmetry of L) such that,

$$\Delta K \subset -x + L,$$

where $\Delta K = \frac{1}{2}K + \frac{1}{2}(-K)$. Hence,

$$W_j(\Delta K) \leq W_j(L),$$

with equality if and only if $\Delta K = -x + L$. But the general Brunn–Minkowski inequality (see e.g. Leichtweiß [**1980**]) gives:

$$W_j(\Delta K) \geq W_j(K),$$

with equality if and only if $K \in \mathcal{K}_s^n$. The desired inequality (11) is just a combination of the last two inequalities.

REFERENCES

1934. T. Bonnesen and W. Fenchel, *Theorie der konvexen Körper*, Springer, Berlin, 1934; English transl., *Theory of Convex Bodies*, BCS Associates, Moscow, Idaho, 1987.

1988. Yu. D. Burago and V. A. Zalgaller, *Geometric Inequalities*, Springer Verlag, Heidelberg, 1988.

1991. H. T. Croft, K. J. Falconer, and R. K. Guy, *Unsolved Problems in Geometry*, Springer, New York, 1991.

1980. K. Leichtweiß, *Konvexe Mengen*, Springer, Berlin, 1980.

1967. C. M. Petty, *Projection bodies*, Proc. Coll. Convexity, Copenhagen, 1965, Københavns Univ. Mat. Inst., 1967, pp. 234–241.

1967. R. Schneider, *Zu einem Problem von Shephard über die Projektionen konvexer Körper*, Math. Z. **101** (1967), 71–82.

1983. R. Schneider and W. Weil, *Zonoids and related topics*, Convexity and its Applications (P. M. Gruber and J. M. Wills, eds.), Birkhäuser, Basel, 1983, pp. 296–317.

1964. G. C. Shephard, *Shadow systems of convex bodies*, Israel J. Math. **2** (1964), 229–236.

DEPARTMENT OF MATHEMATICS, UNIVERSITY OF CALIFORNIA, DAVIS, CA 95616

DEPARTMENT OF MATHEMATICS, POLYTECHNIC UNIVERSITY, BROOKLYN, NY 11201

Contemporary Mathematics
Volume **140**, 1992

Nonlinear Fourier Transform

LEON EHRENPREIS*

I. Introduction

Once, in the early 1960's, some of us were having a discussion as to what directions mathematics would take in the second half of the 20th century. Paul Koosis said "That linear term that appears in the exponential in Fourier analysis will be replaced by a more complicated expression". Certainly the advent of Fourier integral operates has justified Koosis' assertion. In this paper we shall follow a somewhat different path.

Usual (linear) Fourier analysis is most successful in dealing with questions involving

(i) Convex sets or functions defined by inequalities involving exponentials of convex functions.

(ii) Linear partial differential equations with constant coefficients.

(iii) Functions which are small at infinity.

We shall show how more general exponentials can be used to ameliorate the assumption of convexity in (i).

We thought that our approach would lead to an analog of the Fundamental Principle of [2]** for linear partial differential equations with polynomial coefficients in much the same way as linear Fourier transform algebraizes equations with constant coefficients, except that we should add the possibility of solving first order equations, which means, essentially, solving ordinary differential equations. But, up to now, we have not been very successful in this. We can treat only some second order equations, unless we are willing to use infinitely many variables.

As far as (iii) goes, the usual method of dealing with functions that are large, e.g. arbitrary C^∞ functions, is to use the Fourier transform on the dual space,

1991 *Mathematics Subject Classification.* Primary 44A12.

Key words and phrases. Radon transform, Nonlinear Fourier transform.

This paper is in final form and no version of it will be submitted for publication elsewhere

*Supported by a grant from the National Science Foundation

**This work will be referred to as FA.

e.g. the space $\mathcal{E}'$ of distributions of compact support, and then to use duality to define the Fourier transform on $\mathcal{E}$. This works fine for many problems. But sometimes it is difficult to transform the desired properties via duality and Fourier transform.

An (unexpected) bonus of nonlinear Fourier transform is that it has a built in cut-off which allows us to study local properties directly without invoking duality. One of the most striking consequences of this is an analog of Hartogs' separate analyticity theorem that applies to general overdetermined systems of partial differential equations with constant coefficients.

From another point of view the nonlinear Fourier transform of degree 2 of $\Sigma \delta_q$ where q runs over all lattice points in R^n is the theta function. Using the nonlinear Fourier transform of degree > 2 of $\Sigma \delta_q$ leads to higher degree theta functions. Some of their properties are studied in [5].

This paper represents a summary of our ideas. The general theory is set forth in Chapter V of [6]. Details of the application to extensions of solutions will appear in [7].

II. Non convex sets and non convex growth

Around 1970 Paul Malliavin and I studied the question of determining conditions on the Fourier transform $\hat{f}$ of a function $f(x)$ supported in $[-1, 1]$ that determine that f should vanish in a neighborhood of 0. The Paley-Wiener theorem asserts that support $f \subset [-1, 1]$ is equivalent to $\hat{f}$ being an entire function of exponential type 1 with suitable growth on the real axis corresponding to the regularity of f. But the vanishing of f near 0 cannot be described in terms of growth conditions because bounds in x on $exp(ix \cdot \hat{x})$ are satisfied on convex sets (even in R^n with $n > 1$).

Our approach goes as follows: Suppose, for definiteness, that we want to know whether f vanishes on $[-\frac{1}{2}, \frac{1}{2}]$. Let $y = h(x)$ be a convex symmetric curve shaped as in figure 1

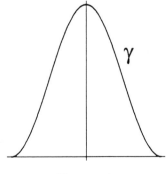

FIGURE 1

Lift the measure $f(x)dx$ to the curve $y = h(x)$. Then the lifted measure $d\mu(x)$ satisfies

(1) $$\text{support } d\mu(x) \subset \{y = h(x)\}.$$

(2) $$\text{support } d\mu(x) \subset \{h(\tfrac{1}{2}) \leq y \leq h(1)\}.$$

Now, let $\hat{\mu}(\hat{x}, \hat{y})$ be the two dimensional Fourier transform of $d\mu$. Moreover, suppose that $h(x)$ is a polynomial. Then properties (1) and (2) translate to

(3) $$\frac{1}{i}\frac{\partial\hat{\mu}}{\partial\hat{y}} = h(\frac{1}{i}\frac{\partial}{\partial\hat{x}})\hat{\mu}$$

(4) $$\exp(ia\hat{y})\hat{\mu}(\hat{x}, \hat{y}) \text{ is of exponential type b in } \hat{y}.$$

Here $a = \frac{1}{2}[h(\frac{1}{2}) + h(1)]$ and $b = \frac{1}{2}[h(1) - h(\frac{1}{2})]$.

In order to relate $\hat{\mu}$ to $\hat{f}$ we observe that

(5) $$\hat{\mu}(\hat{x}, \hat{y}) = \int f(x)e^{ix\hat{x}+ih(x)\hat{y}}dx$$

so that

(6) $$\hat{\mu}(\hat{x}, 0) = \hat{f}(\hat{x}).$$

Thus $\hat{\mu}$ is the solution of the heat-like equation (3) with initial value $\hat{f}$. Condition (4) states that the vanishing of f on $[-\frac{1}{2}, \frac{1}{2}]$ is equivalent to an exponential growth of the solution of this heat-like equation.

Is this condition useful? It is clear that it can be generalized to several variables and to regions defined by polynomial inequalities. The simplest h to use for the question of vanishing near 0 is $h(x) = \Sigma x_j^2$. In our papers [10], [11] Mallivin and I used this idea to study questions of lacunas, that is, when does the fundamental solution e to a hyperbolic equation with constant coefficients have a hole in its support. The above procedure is taylor-made for such problems because $\hat{e}$ can be readily computed by purely algebraic methods.

Instead of dealing with non convex sets we can treat non convex growth conditions. [This contains the case of non convex sets as a special case, if we use the growth condition

(7) $$f(x) = 0(\chi_s(x)$$

where χ_s is the characteristic function of the set S.]

We can study spaces $\mathcal{E}_\psi$ of functions f on R^n satisfying

(8) $$f(x) = 0(e^{-\psi(x)})$$

with similar conditions on derivatives of f and polynomials times f. Instead of assuming that ψ is convex we assume that there is a map

(9) $$H : x \to (h_1(x), \cdots, h_m(x))$$

such that

(10) $\psi(x)$ is the restriction to $\{(x, H(x))\}$ of a convex function ψ_H in R^{n+m}.

The above described method of dealing with support conditions carries over to giving a description of the Fourier transform space $\hat{\mathcal{E}}_\psi$. The analog of the Young conjugate is the Young conjugate of the convex function ψ_H.

Some example of the inverse of this process appear in Chapter V of FA.

III Partial differential equations with polynomial coefficients

We now introduce some notation for the nonlinear Fourier transform on R^n. Denote by

(11) $$p^l = \left(p^l_1, \cdots, p^l_{(l,n)}\right).$$

where (l, n) is the binomial coefficient $\binom{l+n-1}{l}$, an enumeration of the monomials of degree l on R^n. Let $\hat{p}^l$ be a point in $C^{(l,n)}$. Then we define the nonlinear Fourier transform $\hat{f}$ of f of degree l by

(12) $$\hat{f}(\hat{p}^1, \cdots, \hat{p}^l) = \int \exp(i\hat{p}^1 \cdot p^1 + \cdots + i\tilde{\hat{p}}^\ell \cdot p^l) f(x) dx.$$

It is important to observe that $\hat{f}$ satisfies many partial differential equations, for example, the heat-like (or Schroedinger-like) equations

(13) $$\left[p^k_j\left(\frac{1}{i}\frac{\partial}{\partial \hat{p}^1}\right) - \frac{1}{i}\frac{\partial}{\partial \hat{p}^k_j}\right]\hat{f} = 0.$$

for any k, j. Moreover

(14) $$\hat{f}(\hat{p}^1, 0, \cdots, 0) = \hat{f}(\hat{p}^1).$$

Now, suppose that f satisfies some partial differential equation with polynomial coefficients. Any monomial in the coefficients, say p^k_j can be replaced by $\frac{1}{i}\partial/\partial\hat{p}^k_j$ as long as $k \leq l$. If we have a differential operator, say $\partial/\partial x_a$ then the effect on $\hat{f}$ must be calculated by integration by parts. This introduces coefficients of the form $\hat{p}^k_j\hat{p}^{k-1}_{j'}$. Since $k - 1 < l$ the term $p^{k-1}_{j'}$ can be absorbed in a first derivative of $\hat{f}$. This still hods true if $\partial/\partial x_a$ has a linear coefficient. But if it has a quadratic coefficient then a first derivative of $\hat{f}$ will not suffice.

If we had a second order derivative, say $\partial^2/\partial x_a \partial x_b$ then integration by parts would lead to terms like $\hat{p}^k_j\hat{p}^{k'}_{j'}p^{k-1}_{j_1}p^{k'-1}_{j_2}$. Again we can get by with a first order derivative of $\hat{f}$ if $2(l - 1) \leq l$ i.e. $l \leq 2$.

The above argument leads to

PROPOSITION 1. *If ∂ is a linear partial differential operator of order ≤ 2 with polynomial coefficients of degree ≤ 2 such that any first derivatives in ∂ has coefficients of degree ≤ 1 and any second derivative has constant coefficients, then ∂ can be completely resolved by operations involving algebra and the solutions of first order equations.*

Proposition 1 suggests that what is important is the *rank* of a term in ∂, meaning the sum of the degree of the coefficient and the order of the differentiation. Proposition 1 asserts that operators of rank ≤ 2 can be solved in terms of solutions of first order operators. The same idea leads to

PROPOSITION 2. *Operators of rank $\leq 2u$ can be completely resolved by differential operators of order $\leq u$.*

Unfortunately the ranks of the resolving operators of order $\leq u$ may be $2u$ so we cannot iterate the process and end up with first order operators. In order to be able to analyze ∂ using first order operators one has to pass to $l = \infty$. This involves difficulties which we have not yet overcome.

IV Nonlinear Radon transform

In general, the Radon transform deals with the question of recovering a function f from its integrals

$$(15) \qquad \mathbf{R}f(L) = \int_L f$$

Where L is a suitable submanifold of R^n equipped with a fixed measure. The classical Radon transform uses affine L, say all planes of some fixed dimension k. This linear Radon transform can be analyzed in great detail using linear Fourier analysis.

It is important that we put the $\{L\}$ in various classes called <u>spreads</u>. In the linear case a spread consists of all L parallel to a fixed L_o. What is crucial for the translation of the Radon transform to analysis is that a spread consists of all L which form the level sets of some fixed functions $h_1, \cdots, h_{n-k}$.

For the purpose of exposition, let us assume that $n - k = 1$ and that $h = h_1$ is a homogeneous polynomial of degree ℓ. We analyze the geometry using nonlinear Fourier transform of degree ℓ. Since we assume h is homogeneous it suffices to use

$$F(\hat{x}, \hat{p}) = \hat{f}(\hat{x}, 0, \cdots, 0, \hat{p})$$
$$(16) \qquad = \int f(x) e^{i\hat{x} \cdot x + i\hat{p} \cdot p^l} \, dx.$$

We can write $h(x) = \hat{p}^0 \cdot p^\ell$ where $\hat{p}^0$ is some (real) point in $R^{(\ell,n)}$. For any scalar t we write

$$F(0, t\hat{p}^o) = \int f(x) e^{it\hat{p}^o \cdot p^l} dx$$

$$= \int e^{itc} dc \int_{\hat{p}^o \cdot p^l = c} f(x) dx$$

(17)
$$= \int e^{itc} \mathbf{R} f(\hat{p}^o \cdot p^l = c) dc$$

if we use as measure on $\hat{p}^o \cdot p^l = c$ the euclidean measure divided by $| \operatorname{grad} \hat{p}^o \cdot p^l |$. Since ordinary Fourier transform is an isomorphism we conclude

PROPOSITION 3. *The geometric information conveyed by* $\mathbf{R}f$ *on the spread* $\{\hat{p}^o \cdot p^l = c\}$ *is equivalent to the analytic information* $\{ F(0, \hat{p})|_{\hat{p}=t\hat{p}^o} \}$.

Recall that $\hat{f}$ satisfies many partial differential equations. They imply equations for $F(0, \hat{p})$ on the $\hat{p}$ axis; we call these Plücker equations. The geometry tells us about $F(0, \hat{p})$ on the set of lines corresponding to the geometric spreads. The Plücker equations tell us that we need a set of lines forming a uniqueness set for solutions of the Plücker equations in order to obtain an injective Radon transform.

As the algebra and analysis are quite involved, let us illustrate the case $l = 2, n = 2$. We set

$$\hat{p}_1 \cdot p^2 = x_1^2$$
$$\hat{p}_2 \cdot p^2 = x_2^2$$
(18)
$$\hat{p}_3 \cdot p^2 = x_1 x_2$$

(Note that 2 is a superscript for p^2 and a power for x_1^2, x_2^2.) The Plücker equation is the wave equation

(19)
$$\left[\frac{\partial^2}{\partial \hat{p}_1 \partial \hat{p}_2} - \frac{\partial^2}{\partial \hat{p}_3^2} \right] F(0, \hat{p}) = 0.$$

We need enough spreads to provide a uniqueness set for the wave equation. It is not difficult to add derivatives to the geometry, so the simplest situation would involve lines filling the Cauchy surface

(20)
$$\hat{p}_1 + \hat{p}_2 = 0.$$

One is then left with the problem of determining $F(\hat{x}, \hat{p})$ from $F(0, \hat{p})$ since eventually we are really interested in $F(\hat{x}, 0) = \hat{f}(\hat{x})$. To determine $F(\hat{x}, \hat{p})$ we use the heat-like equations (13). The $\hat{p}$ axis is an _exotic Cauchy surface_ for this system, meaning that, while there is a good Cauchy Problem with data on the $\hat{p}$ axis, this data is restricted to satisfy various equations including the Plücker equations. In fact, we cannot assign "independent data" on the $\hat{p}$ axis.

A detailed study of the Radon transform and of exotic Cauchy Problems is found in my book [6].

V Cut-offs and extension of solutions of partial differential equations

In Section II we explained how the nonlinear Fourier transform can be used to determine the exact support of a function. How about various notions of singular support or wave front sets?

Let us start with $n = 1$. Let f be defined in a neighborhood of $[-1, 1]$. We want to know if f can be extended to be holomorphic in neighborhood of 0 in the upper half plane.

If f were defined on the whole real axis and small at infinity, and if we wanted the extension to be holomorphic and small at infinity in a whole strip $[0 \leq \Re x \leq \epsilon]$ then we could settle this problem by linear Fourier analysis.

Actually if f is only defined locally we cannot even take its nonlinear Fourier transform in any reasonable sense. However, we can cut f by multiplying by a smooth function which is 1 near the origin. We want to show that nonlinear Fourier transform allows us to ignore the cut-off and thus deal with the local problem as though it were a global problem.

Let us examine $\hat{\hat{f}}(\hat{p}^1, \cdots, \hat{p}^l)$. Since f has been cut-off $\hat{\hat{f}}$ is an entire function. To make use of the analyticity of f we shift the contour in the integral defining $\hat{\hat{f}}$ to the contour γ in figure 2:

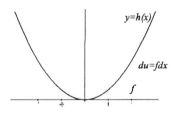

FIGURE 2

Now, make use of the fact that when $\Im x > 0$, $\Im i\hat{p}^1 x < 0$ when $\hat{p}^1 > 0$. Thus the term $\exp(i\hat{p}^1 x)$ is exponentially decreasing in $\hat{p}^1$ as $\hat{p}^1 \to +\infty$ when we are near the top of γ. But other terms, e.g. $\exp(i\hat{p}^2 x^2)$ in the nonlinear exponential are exponentially decreasing in $\hat{p}^2$ near the bottom of γ when $\Im\hat{p}^2 \to +\infty$. (Note that 2 is an exponent for x^2 and 2 is a superscript for $\hat{p}^2$.) However, we do not get exponential decrease simultaneously in $\hat{p}^1, \hat{p}^2$ because $\Im\hat{p}^2 x^2$ becomes negative on the top of γ when $\Im\hat{p}^2 > 0$. Thus we must balance $\hat{p}^1$ and $\hat{p}^2$ to obtain exponential decrease. This balance can be achieved in a suitable sector of the quadrant in the plane where $\hat{p}^1$ is real, $\hat{p}^2$ pure imaginary and $\hat{p}^1 > 0$,

A similar analysis works for $\hat{p}^j$ with $j \geq 2$.

Once we have the exponential decrease we want to get back and show that f has an extension to be holomorphic in the neighborhood of 0 in $\Im x > 0$. For this we use the heat-like equations. We express f in terms of its Fourier transform $\hat{f}$.

The heat-like equations enable us, via Green's theorem (integration by parts), to shift the contour to the region of exponential decrease. This verifies the analytic extension of f.

In case we restrict ourselves to $l = 2$ this method yields about the same results as those obtained from the FBI transform of Bros and Iagolnitzer [1]. But by use of general l T. Banh succeeded in determining the exact domain of analyticity of f in terms of the nonlinear Fourier transform.

Once we have a characterization of the Fourier transform of such f, we can put things together. For example if f extends to be holomorphic both in $\Im x > 0$ and $\Im x < 0$ near 0 then f extends to be holomorphic in a full neighborhood of 0. This result is, of course, trivial. But it points the way to the nontrivial (though still well-known) analog in higher dimension: Suppose $n > 0$ and for any $k \leq n$ for any $x_1, \cdots, x_{k-1}, x_{k+1}, \cdots x_n$ in some fixed real neighborhood of 0 the function f extends to be holomorphic in a fixed neighborhood of the origin in the complex half plane $\Im x_k > 0$. Then f extends to a holomorphic function in the complex region

$$|x| < \epsilon$$

(21)
$$\Im x_j > 0 \text{ for all } j.$$

This theorem is remarkable because f starts off being holomorphic on sets of real dimension $n + 1$ and ends up being holomorphic on a set of real dimension $2n$. It is due to S. Bernstein and Hartogs.

The same idea applies to the edge-of-the-wedge theorem. With some modification we can study C^∞ wave front sets and Denjoy-Carleman wave front sets.

We again emphasize that all these results are easily derived by methods of linear Fourier transform when everything is globally defined and small at infinity. Thus for many problems the usage of the nonlinear terms allows us to treat local problems.

We want to reformulate the Bernstein-Hartogs result so as to put it in the framework of systems of partial differential equations. Thus "holomorphicity" should be generalized to "solution of a general overdetermined system of linear partial differential equations". To simplify our exposition we assume that the operators have constant coefficients. We set

(22)
$$\vec{P}(D) = (\vec{P}_1(D), \cdots, \vec{P}_r(D))$$

and we write

(23)
$$\vec{P}G = 0$$

to mean $P_j(D)G = 0$ for all j.

In the Bernstein-Hartogs theorem we started with a function f on R^n. From the point of view of general partial differential equations, R^n is a Lagrangian subspace for $\bar{\partial}$ meaning that it is a maximal manifold on which the restrictions of solutions of $\bar{\partial}$ do not satisfy any differential equations. Putting things another way, there is no operator η of the form

$$(24) \qquad\qquad \eta = \Sigma \lambda_j \frac{\partial}{\partial \bar{z}_j}$$

on C^n which lives on R^n meaning that $\eta g = 0$ on R^n if $g = 0$ on R^n. (Or, what is the same thing, η restricted to R^n can be expressed in terms of tangential derivatives to R^n.) Here λ_j are arbitrary differential operators.

For our general $\vec{P}$ we term $\vec{P}$ <u>Lagrangian</u> a maximal submanifold L such that no operator in the left ideal generated by $\vec{P}$ can be expressed, on L, in terms of tangential derivatives. We shall restrict our considerations to linear L. In this case L is a $\vec{P}$ Lagrangian if it is maximal amongst linear spaces with the property that the projection on the complexification of L of the complex variety V of common zeros of the $P_j(ix)$ is a Zariski open set in V. Thus dim $L = $ dim $V = n$. We call N the dimension of x so that $N = 2n$ for $\bar{\partial}$.)

It follows from the results of Chapter IX of FA that a general $\vec{P}$ Lagrangian L can be used for a non exotic Cauchy surface for the system $\vec{P}$. Thus L takes the place of R^n for $\bar{\partial}$ and the Cauchy data $\vec{f}$ takes the place of the function f on R^n.

What corresponds to the hypothesis on the extendibility of $f(x)$ to complex x_k when $x_1, \cdots, x_{k-1}, x_{k+1}, \cdots, x_n$ are fixed? Let v be a vector not in L and call

$$(25) \qquad\qquad L_v = \text{linear space generated by } L \text{ and } v.$$

We know that $\vec{P}$ induces an equation $\vec{P}_v$ on L_v because L is a Lagrangian. This corresponds to the holomorphic extendibility condition on f.

Our main result is

THEOREM. *Suppose that for a suitably chosen finite set $v_1, \cdots, v_{N-n}$ the proposed Cauchy Data $\hat{f}$ is actually (in a suitable sense) the Cauchy data on L of a solution F_j of $\vec{P}_{v_j} F_j = 0$. Here F_j defined and is a solution of $\vec{P}_{v_j}$ on a fixed neighborhood of zero in $L_{v_j} \cap \{tv_j | t \geq 0\}$. Then there is a solution F of $\vec{P}F = 0$ defined on a whole local neighborhood of zero in R^N intersected with $\underset{j}{\cap}\{tv_j | t \geq 0\}$*

The proof of the theorem follows along the same general lines as the proof of the Bernstein-Hartogs theorem outlined above. Shifting the contour in x is done using a suitable version of integration by parts. However the nonlinear terms are defined by forcing them to be solutions of heat-like equations rather than by the simpler algebraic means that we applied to $\bar{\partial}$.

The details of the proof will be published elsewhere.

REFERENCES

1. J. Bros and D. Iagolnitzer, *support essentiel et structure analytiques des distributions*, Séminaire Goulaovic–Lions–Schwartz (1975-1976 No. 18).
2. L. Ehrenpreis, *Fourier Analysis in Several Complex Variables*, Wiley, Interscience, New York (1970).

3. _____, *Complex Fourier transform technique in variable coefficient partial differential equations*, J.d'Analyse (Jerusalem) **XIX** (1967), 75-95.

4. _____, *Lewy unsolvability and several complex variables*, Mich. Math. J. **38(1991)**, 417-439.

5. _____, *Fourier analysis, partial differential equations, and automorphic functions in Theta Functions, Bowdain 1987*, Proc. Symposia Pure M. **49(2)**, 45-100.

6. _____, *The Radon Transform (to appear)*.

7. _____, *Extensions of solutions of partial differential equation, in Geometric and Algebraical Aspects of Several Complex Variables*, Cetraro (Italy) (1992).

8. _____, *The Rogers-Ramanujan identities*, to appear in Emil Grosswald Memorial Volume (Springer).

9. _____, *Partial differential equations and nonlinear Fourier analysis*, in preparation.

10. _____, *Fourier analysis on nonconvex sets*, in Symposia Mat. Instituto Nazionale de Alta Mat, **XII** (1968).

11. _____, *Spectral gaps and lacunas*, Bull. Sc. Mat. **105** (1970), 17-28.

DEPARTMENT OF MATHEMATICS, TEMPLE UNIVERSITY, BROAD & MONTGOMERY, PHILADELPHIA, PA 19122

Contemporary Mathematics
Volume **140**, 1992

On the Infinitesimal Rigidity of the Complex Quadrics

HUBERT GOLDSCHMIDT

ABSTRACT. We present outlines of our proofs of the infinitesimal rigidity of the complex quadric of dimension ≥ 4 in conjunction with Michel's work on the real projective plane.

Introduction

A symmetric 2-form h on a compact symmetric space (X, g) satisfies the zero-energy condition if for all closed geodesics γ of X the integral

$$\int_\gamma h = \int_0^L h(\dot\gamma(s), \dot\gamma(s)) \, ds$$

of h over γ vanishes, where $\dot\gamma(s)$ is the tangent vector to the geodesic γ parametrized by its arc-length s and L is the length of γ. We denote by $F_2(X)$ the space of all symmetric 2-forms on X satisfying the zero-energy condition. A Lie derivative of the metric g along a vector field always satisfies the zero-energy condition. The space (X, g) is infinitesimally rigid if the only symmetric 2-forms on X satisfying the zero-energy condition are the Lie derivatives of the metric g.

The infinitesimal rigidity property can be viewed as a generalization of the injectivity of the Radon transform. Indeed, if f is a function on X, its Radon transform $\hat{f}$ is a function on the set of all closed geodesics of X whose value at γ is

$$\hat{f}(\gamma) = \int_\gamma f.$$

The injectivity of the Radon transform has been established for all compact symmetric spaces, which are not spheres (see [1] for the rank 1 case and [14] for higher rank).

1991 *Mathematics Subject Classification*. Primary 53C35, 53C65.

This paper is in final form and no version of it will be submitted for publication elsewhere

Determining whether a space is infinitesimally rigid is an infinitesimal version of the Blaschke Conjecture. Indeed, if (X, g) is a projective space, different from a sphere, with its canonical metric g, all of its geodesics are closed and of the same length π. The Blaschke Conjecture asserts that any other metric on X with same property is isometric to g. If $\{g_t\}$ is a deformation of this metric $g_0 = g$ and if g_t has this property for all t, then the infinitesimal deformation $\frac{d}{dt} g_t|_{t=0}$ satisfies the zero-energy condition.

The infinitesimal rigidity of these spaces was established by R. Michel [13] for the real projective spaces $\mathbb{RP}^n$, and by R. Michel [13] and C. Tsukamoto [17] for the other projective spaces (see also [5] and [7]). On the other hand, R. Michel [14] demonstrated the infinitesimal rigidity of flat tori.

In [7], [8] and [9], with J. Gasqui, we showed that the complex quadric of dimension ≥ 4 is infinitesimally rigid. This is the first contribution to the rigidity question for irreducible symmetric spaces of rank > 1.

Here we present outlines of our proofs of the infinitesimal rigidity of the quadric. We point out how our methods used in the study [7] of the quadric of dimension ≥ 5 break down for the quadric of dimension four.

For our analysis of the quadric of dimension four, we introduce a new approach to infinitesimal rigidity partially inspired by Michel's work [13]. In §2, we give both Michel's original proof of the infinitesimal rigidity of the real projective plane $\mathbb{RP}^2$ and the variant due to Bourguignon (see [1]). We attempt to show how our arguments for the quadric of dimension four correspond to various assertions proved in §2 for $\mathbb{RP}^2$. Our techniques rely on results about differential operators on symmetric spaces (see §1) and on harmonic analysis on homogeneous spaces. In our approach, the Lichnerowicz Laplacian acting on symmetric 2-forms plays a crucial role. We try to explain why this is the case here and in Michel's work [13], where its use often seemed so ad hoc. The complete proofs can be found in [8] and [9].

1. Locally symmetric Einstein manifolds

Let (X, g) be a Riemannian manifold of dimension m. We denote by T and T^* its tangent and cotangent bundles. By $\bigotimes^k E$, $S^l E$, $\bigwedge^j E$, we shall mean the k-th tensor product, the l-th symmetric product and the j-th exterior product of a vector bundle E over X, respectively. If E is a vector bundle over X, we denote by $E_\mathbb{C}$ its complexification, by $\mathcal{E}$ the sheaf of sections of E over X and by $C^\infty(E)$ the space of global sections of E over X. If E, F are vector bundles over X and $P : \mathcal{E} \to \mathcal{F}$ is a differential operator, we write $\operatorname{Ker} P$ for the kernel of $P : C^\infty(E) \to C^\infty(F)$. Let $C^\infty(X)$ be the space of real-valued functions on X. If ξ is a vector field on X and β is a section of $\bigotimes^k T^*$ over X, we denote by $\mathcal{L}_\xi \beta$ the Lie derivative of β along ξ. Let $g^\flat : T \to T^*$, $g^\sharp : T^* \to T$ be the isomorphisms determined by the metric g.

Let G be the sub-bundle of $\bigwedge^2 T^* \otimes \bigwedge^2 T^*$ consisting of those tensors satisfying the first Bianchi identity considered in [**5**, §3]. Let

$$\text{Tr} : S^2 T^* \to \mathbb{R}, \qquad \text{Tr} : G \to S^2 T^*$$

be the the trace mappings defined by

$$\text{Tr}\, h = \sum_{j=1}^{m} h(t_j, t_j), \qquad (\text{Tr}\, u)(\xi, \eta) = \sum_{j=1}^{m} u(t_j, \xi, t_j, \eta),$$

for $h \in S^2 T_x^*$, $u \in G_x$, where $x \in X$ and $\{t_1, \dots, t_m\}$ is an orthonormal basis of T_x. We denote by G^0 the kernel of $\text{Tr} : G \to S^2 T^*$.

We now introduce various differential operators associated to the Riemannian manifold (X, g). First, let ∇ be the Levi-Civita connection of (X, g). The Killing operator

$$D_0 : T \to S^2 T^*$$

of (X, g), sending $\xi \in T$ into $\mathcal{L}_\xi g$, and the symmetrized covariant derivative

$$D^1 : T^* \to S^2 T^*$$

defined by

$$(D^1 \theta)(\xi, \eta) = \tfrac{1}{2}((\nabla \theta)(\xi, \eta) + (\nabla \theta)(\eta, \xi)),$$

for $\theta \in T^*$, $\xi, \eta \in T$, are related by the formula

$$\tfrac{1}{2} D_0 \xi = D^1 g^\flat(\xi),$$

for $\xi \in T$. Consider the first-order differential operator

$$\text{div} : S^2 T^* \to T^*$$

and the Laplacian

$$\overline{\Delta} : S^2 T^* \to S^2 T^*$$

defined by

$$(\text{div}\, h)(\xi) = -\sum_{j=1}^{m} (\nabla h)(t_j, t_j, \xi), \qquad (\overline{\Delta} h)(\xi, \eta) = -\sum_{j=1}^{m} (\nabla^2 h)(t_j, t_j, \xi, \eta),$$

for $h \in C^\infty(S^2 T^*)$, $\xi, \eta \in T_x$, where $x \in X$ and $\{t_1, \dots, t_m\}$ is an orthonormal basis of T_x. The formal adjoint of D_0 is equal to $2g^\sharp \cdot \text{div} : S^2 T^* \to T$. Since D_0 is elliptic, if X is compact, we therefore have the orthogonal decomposition

(1.1) $$C^\infty(S^2 T^*) = D_0 C^\infty(T) \oplus \text{Ker div}$$

(see [**3**]).

Let $\mathcal{R}(h)$ and $\text{Ric}(h)$ be the Riemann curvature tensor, as defined in [**5**, §4], and the Ricci tensor of a metric h on X, which is are sections of G and $S^2 T^*$, respectively. We set $R = \mathcal{R}(g)$ and $\text{Ric} = \text{Ric}(g)$; we have $\text{Ric} = -\text{Tr}\, R$. Let $\widetilde{R}$ be the section of $\bigwedge^2 T^* \otimes T^* \otimes T$ determined by

$$g(\widetilde{R}(\xi_1, \xi_2, \xi_3), \xi_4) = R(\xi_1, \xi_2, \xi_3, \xi_4),$$

for $\xi_1, \xi_2, \xi_3, \xi_4 \in T$. Let

$$\mathcal{R}'_g : S^2 T^* \to \mathcal{G}, \qquad \mathrm{Ric}'_g : S^2 T^* \to S^2 T^*$$

be the linear differential operators of order 2 which are the linearizations along g of the non-linear operators $h \mapsto \mathcal{R}(h)$ and $h \mapsto \mathrm{Ric}(h)$ respectively, where h is a Riemannian metric on X. If $h \in C^\infty(S^2 T^*)$ and $h(x) = 0$, with $x \in X$, then we have

$$(1.2) \qquad \mathrm{Tr}\,(\mathcal{R}'_g h)(x) = -(\mathrm{Ric}'_g h)(x).$$

If Δ is the Lichnerowicz Laplacian of [**12**](for an intrinsic definition of Δ, see [**10**, §4]), we have

$$(1.3) \qquad \mathrm{Tr}\,\Delta h = \Delta \mathrm{Tr}\, h,$$

for $h \in S^2 T^*$. If (X, g) has constant curvature K, then we have

$$(1.4) \qquad \Delta h = \overline{\Delta} h + 2mKh - 2K\mathrm{Tr}\, h \cdot g,$$

for $h \in S^2 T^*$. The operator Ric'_g is expressed in terms of the Lichnerowicz Laplacian; in fact, we have

$$(1.5) \qquad \mathrm{Ric}'_g h = \tfrac{1}{2}\Delta h - D^1 \mathrm{div}\, h - \tfrac{1}{2}\mathrm{Hess}\,\mathrm{Tr}\, h,$$

for $h \in S^2 T^*$ (see for example [**2**, Theorem 1.174]). Thus if for $h \in S^2 T^*$ satisfies $\mathrm{div}\, h = 0$, then by (1.3) we see that

$$(1.6) \qquad \mathrm{Ric}'_g h = \tfrac{1}{2}\Delta h - \tfrac{1}{2}\mathrm{Hess}\,\mathrm{Tr}\, h,$$

$$(1.7) \qquad \mathrm{Tr}\,\mathrm{Ric}'_g h = \Delta \mathrm{Tr}\, h.$$

LEMMA 1.1. *If (X, g) is a two-dimensional manifold of constant curvature K and h is a symmetric 2-form on an open subset U of X satisfying $\mathrm{div}\, h = 0$ and $\mathrm{Tr}\, h = 0$, then we have*

$$\overline{\Delta} h + 2Kh = 0, \qquad \Delta h = 2Kh$$

on U.

PROOF. Let $x \in X$ and $\{\xi_1, \xi_2\}$ be an orthonormal basis of T_x. The condition $\mathrm{div}\, h = 0$ implies the following identities

$$(\nabla^2 h)(\xi_1, \xi_1, \xi_1, \xi_1) + (\nabla^2 h)(\xi_1, \xi_2, \xi_2, \xi_1) = 0,$$

$$(\nabla^2 h)(\xi_2, \xi_1, \xi_1, \xi_2) + (\nabla^2 h)(\xi_2, \xi_2, \xi_2, \xi_2) = 0;$$

from the condition $\mathrm{Tr}\, h = 0$, we obtain

$$(\nabla^2 h)(\xi_1, \xi_1, \xi_1, \xi_1) + (\nabla^2 h)(\xi_1, \xi_1, \xi_2, \xi_2) = 0.$$

From all these equalities, we infer that

$$(\overline{\Delta}h)(\xi_2, \xi_2) = -(\nabla^2 h)(\xi_1, \xi_2, \xi_2, \xi_1) + (\nabla^2 h)(\xi_2, \xi_1, \xi_1, \xi_2)$$
$$= h(\widetilde{R}(\xi_1, \xi_2)\xi_1, \xi_2) + h(\xi_1, \widetilde{R}(\xi_1, \xi_2)\xi_2)$$
$$= K(h(\xi_1, \xi_1) - h(\xi_2, \xi_2)),$$

and, since $\operatorname{Tr} h = 0$, that

$$(\overline{\Delta} + 2Kh)(\xi_2, \xi_2) = 0.$$

By polarization, we see that $\overline{\Delta}h + 2Kh = 0$. By (1.4), we thus have $\overline{\Delta}h = 2Kh$.

Let $\widetilde{G}$ be the sub-bundle of G, with variable fiber, whose fiber at $x \in X$ is the infinitesimal orbit of the curvature

$$\widetilde{G}_x = \{ (\mathcal{L}_\xi R)(x) \mid \xi \in T_x \text{ with } (\mathcal{L}_\xi g)(x) = 0 \},$$

and let $\alpha : G \to G/\widetilde{G}$ be the canonical projection. If $\widetilde{G}$ is a vector bundle, we consider the second-order differential operator

$$D_1 : S^2 T^* \to \mathcal{G}/\widetilde{\mathcal{G}}$$

introduced in [5] and determined by

$$(D_1 h)(x) = \alpha(\mathcal{R}'_g(h - \mathcal{L}_\xi g))(x),$$

for $x \in X$ and $h \in S^2 T^*_x$, where ξ is an element of T_x satisfying $h(x) = (\mathcal{L}_\xi g)(x)$. It is easily seen that this operator is well-defined and that $D_1 \cdot D_0 = 0$.

If (X, g) has constant curvature, then $\widetilde{G} = 0$ and the sequence

$$T \xrightarrow{D_0} S^2 T^* \xrightarrow{D_1} \mathcal{G}$$

is exact; moreover, if X is simply-connected, the sequence

$$(1.8) \qquad C^\infty(T) \xrightarrow{D_0} C^\infty(S^2 T^*) \xrightarrow{D_1} C^\infty(G)$$

is exact (see [5, Theorem 6.1]). Thus if X is the sphere S^n, or a quotient of S^n by a finite group of isometries, the sequence (1.8) is exact. In particular, (1.8) is exact when X is the real projective space $\mathbb{RP}^n$.

If X is the complex projective space $\mathbb{CP}^n$, with $n \geq 2$, or the complex quadric Q_n of dimension n (see §3), with $n \geq 3$, then $\widetilde{G}$ is a vector bundle and the sequences

$$T \xrightarrow{D_0} S^2 T^* \xrightarrow{D_1} \mathcal{G}/\widetilde{\mathcal{G}},$$

$$(1.9) \qquad C^\infty(T) \xrightarrow{D_0} C^\infty(S^2 T^*) \xrightarrow{D_1} C^\infty(G/\widetilde{G})$$

are exact (see [5, §7] and [7, §5]); thus D_1 is the compatibility condition for the operator D_0.

Throughout the remainder of this paper, we assume that (X, g) is a connected locally symmetric Einstein manifold; we have $\operatorname{Ric} = \lambda g$, with $\lambda \in \mathbb{R}$. Then $\widetilde{G}$ is a vector bundle over X.

LEMMA 1.2. *We have*

$$\widetilde{G} \subset G^0.$$

According to Lemma 1.2, the mapping $\mathrm{Tr}: G \to S^2T^*$ induces by passage to the quotient a morphism of vector bundles

$$\mathrm{Tr}: G/\widetilde{G} \to S^2T^*.$$

In [**9**, §1], using (1.2) we verify

PROPOSITION 1.2. *For $h \in S^2T^*$, we have*

$$-\mathrm{Tr}\, D_1 h = \mathrm{Ric}'_g h - \lambda h.$$

If (X, g) is a surface of constant curvature K, then $\mathrm{Ric} = Kg$; moreover G is a line bundle and so

$$\mathrm{Tr} \cdot \mathrm{Tr}: G_x \to \mathbb{R}$$

is an isomorphism for all $x \in X$. Therefore, according to Proposition 1.2, (1.5) and (1.3), D_1 is determined by the relation

$$(1.10) \qquad -\mathrm{Tr} \cdot \mathrm{Tr}\, D_1 h = \Delta \mathrm{Tr}\, h - K \mathrm{Tr}\, h - \mathrm{Tr}\, D^1 \mathrm{div}\, h.$$

Now suppose that X is the real projective plane $\mathbb{RP}^2$ with its metric of constant curvature 1. Let h be an element of $C^\infty(S^2T^*)$ satisfying $\mathrm{div}\, h = 0$. Then by (1.10), we see that $D_1 h = 0$ if and only if $\Delta \mathrm{Tr}\, h = \mathrm{Tr}\, h$. Since the first eigenvalue of $\mathbb{RP}^2$ is equal to 6, the relation $D_1 h = 0$ is equivalent to $\mathrm{Tr}\, h = 0$. Thus we obtain

$$(1.11) \qquad \mathrm{Ker}\, D_1 \cap \mathrm{Ker}\, \mathrm{div} = \{\, h \in C^\infty(S^2T^*) \mid \mathrm{div}\, h = 0,\ \mathrm{Tr}\, h = 0 \,\},$$

and so by (1.1) we see that $\mathrm{Ker}\, D_1 \cap \mathrm{Ker}\, \mathrm{div}$ is the orthogonal complement of $D_0 C^\infty(T) + C^\infty(X)g$ in $C^\infty(S^2T^*)$. Since the sequence (1.8) is exact, we know that

$$(1.12) \qquad \mathrm{Ker}\, D_1 \cap \mathrm{Ker}\, \mathrm{div} = 0;$$

we now deduce that

$$(1.13) \qquad C^\infty(S^2T^*) = D_0 C^\infty(T) + C^\infty(X)g.$$

2. The real projective plane

We suppose that (X, g) is the real projective plane $\mathbb{RP}^2$, with its canonical metric of constant curvature 1. Let $\gamma: [-\pi/2, \pi/2] \to X$ be a closed geodesic parametrized by its arc-length. We set $\gamma(0) = x$, and let $e_1(t) = \dot\gamma(t)$ be the tangent vector to the geodesic at $\gamma(t)$, for $-\pi/2 \le t \le \pi/2$. We choose a unit vector $e_2 \in T_x$ orthogonal to $e_1(0)$, and consider the family of tangent vectors $e_2(t) \in T_{\gamma(t)}$, with $-\pi/2 \le t \le \pi/2$, obtained by parallel transport of the vector

e_2 along γ. Clearly, if θ is an element of $C^\infty(\bigotimes^p T^*)$ and $\xi_j(t)$ is a vector field along $\gamma(t)$ equal to either $e_1(t)$ or $e_2(t)$ for $1 \le j \le p-1$, we have

$$(\nabla\theta)(e_1(t), \xi_1(t), \ldots, \xi_{p-1}(t)) = \frac{d}{dt}\theta(\xi_1(t), \ldots, \xi_{p-1}(t)),$$

and so

(2.1)
$$\int_{-\frac{\pi}{2}}^{\frac{\pi}{2}} (\nabla\theta)(e_1(t), \xi_1(t), \ldots, \xi_{p-1}(t))\, dt = 0.$$

The vector fields

$$J_1(t) = \sin t \cdot e_2(t), \qquad J_2(t) = \cos t \cdot e_2(t)$$

along γ are a basis for the space of normal Jacobi fields along γ. For $i = 1$, 2, let

$$\Phi_i : [-\pi/2, \pi/2] \times (-\varepsilon, \varepsilon) \to X$$

be the geodesic variation of γ corresponding to the Jacobi field J_i. For $-\varepsilon < s < \varepsilon$, let

$$\gamma_{i,s} : [-\pi/2, \pi/2] \to X$$

be the closed geodesic defined by

$$\gamma_{i,s}(t) = \Phi_i(t, s),$$

for $-\pi/2 \le t \le \pi/2$.

Let h be an element of $C^\infty(S^2 T^*)$. We consider the real-valued functions $E_1(h)$, $E_2(h)$ defined on the neighborhood $(-\varepsilon, \varepsilon)$ of 0 by

$$E_i(h)(s) = \int_{\gamma_{i,s}} h,$$

for $i = 1$, 2 and $-\varepsilon < s < \varepsilon$. In [13], Michel shows that

$$\frac{d^2 E_1(h)}{ds^2}(0) + \frac{d^2 E_2(h)}{ds^2}(0) = \int_{-\frac{\pi}{2}}^{\frac{\pi}{2}} \{(\nabla^2 h)(e_2(t), e_2(t), e_1(t), e_1(t))$$

$$+ 2(h(e_2(t), e_2(t)) - h(e_1(t), e_1(t)))\}\, dt$$

$$= -\int_{-\frac{\pi}{2}}^{\frac{\pi}{2}} (\Delta h)(e_1(t), e_1(t))\, dt.$$

The second equality is a direct consequence of (1.4) and (2.1).

LEMMA 2.1. *The space $F_2(\mathbb{RP}^2)$ of symmetric 2-forms satisfying the zero-energy condition is stable under the Lichnerowicz Laplacian.*

PROOF. For $h \in F_2(\mathbb{RP}^2)$, the energy functions $E_1(h)$ and $E_2(h)$ vanish identically. By the preceding equality, we obtain

$$\int_\gamma \Delta\theta = 0.$$

LEMMA 2.2. *If $h \in F_2(\mathbb{RP}^2)$ satisfies $\operatorname{div} h = 0$, then we have*

$$\int_{-\frac{\pi}{2}}^{\frac{\pi}{2}} (\nabla^2 h)(e_2(t), e_2(t), e_2(t), e_2(t))\, dt$$

$$= \int_{-\frac{\pi}{2}}^{\frac{\pi}{2}} h(e_2(t), e_2(t))\, dt = \int_{-\frac{\pi}{2}}^{\frac{\pi}{2}} (\operatorname{Tr} h)(\gamma(t))\, dt.$$

PROOF. We have

$$(\nabla^2 h)(e_2(t), e_2(t), e_2(t), e_2(t)) = -(\nabla^2 h)(e_2(t), e_1(t), e_1(t), e_2(t))$$

$$= -(\nabla^2 h)(e_1(t), e_2(t), e_1(t), e_2(t))$$

$$+ h(e_2(t), e_2(t)) - h(e_1(t), e_1(t));$$

the first equality holds because $\operatorname{div} h = 0$, while the second one is obtained using the expression for the curvature of (X, g). The lemma is now a consequence of (2.1).

LEMMA 2.3. *If $h \in F_2(\mathbb{RP}^2)$ satisfies $\operatorname{div} h = 0$, then*

$$\int_{\gamma} (\Delta \operatorname{Tr} h - \operatorname{Tr} h) = 0.$$

PROOF. According to (1.4), we have

$$(\Delta h)(e_2(t), e_2(t))$$

$$= -(\nabla^2 h)(e_1(t), e_1(t), e_2(t), e_2(t)) - (\nabla^2 h)(e_2(t), e_2(t), e_2(t), e_2(t))$$

$$+ 4h(e_2(t), e_2(t)) - 2(\operatorname{Tr} h)(\gamma(t)).$$

By (1.2), (2.1), Lemmas 2.1 and 2.2, and the preceding equality, we see that

$$\int_{-\frac{\pi}{2}}^{\frac{\pi}{2}} (\Delta \operatorname{Tr} h)(\gamma(t))\, dt = \int_{-\frac{\pi}{2}}^{\frac{\pi}{2}} (\operatorname{Tr} \Delta h)(\gamma(t))\, dt = \int_{-\frac{\pi}{2}}^{\frac{\pi}{2}} (\Delta h)(e_2(t), e_2(t))\, dt$$

$$= \int_{-\frac{\pi}{2}}^{\frac{\pi}{2}} (\operatorname{Tr} h)(\gamma(t))\, dt.$$

LEMMA 2.4. *If $h \in F_2(\mathbb{RP}^2)$ satisfies $\operatorname{div} h = 0$, then $\operatorname{Tr} h$ vanishes.*

PROOF. Since the Radon transform on $\mathbb{RP}^2$ is injective, from Lemma 2.3 we deduce that $\Delta \operatorname{Tr} h = \operatorname{Tr} h$. As the first eigenvalue of $\mathbb{RP}^2$ is equal to 6, we see that $\operatorname{Tr} h = 0$.

LEMMA 2.5. *An element h of $F_2(\mathbb{RP}^2)$ satisfying $\operatorname{div} h = 0$ vanishes.*

PROOF. According to Lemmas 2.4 and 1.1, we know that $\overline{\Delta} h + 2h = 0$. Since $\overline{\Delta}$ is a positive operator, we see that $h = 0$.

Alternatively, Lemma 2.5 can also be obtained using the operator D_1 and the exact sequence (1.8). Indeed, if h is an element of $F_2(\mathbb{RP}^2)$ satisfying $\operatorname{div} h = 0$, according to Lemma 2.4 we have $\operatorname{Tr} h = 0$, and so by (1.11) we know that $D_1 h = 0$. From (1.12), it now follows that $h = 0$.

THEOREM 2.1. *The real projective plane* $\mathbb{RP}^2$ *is infinitesimally rigid.*

PROOF. Let h be an element of $F_2(\mathbb{RP}^2)$; according to (1.1), we may write $h = \mathcal{L}_\xi g + h'$, where ξ is a vector field on X and h' is an element of $C^\infty(S^2 T^*)$ satisfying div $h' = 0$. Since $\mathcal{L}_\xi g$ belongs to $F_2(\mathbb{RP}^2)$, so does h'. From Lemma 2.5, we deduce that $h' = 0$.

We now present Bourguignon's proof of Theorem 2.1 given in [1]. Let h be an element of $F_2(\mathbb{RP}^2)$. According to (1.3), we may write

$$h = \mathcal{L}_\xi g + fg,$$

where ξ is a vector field on X and $f \in C^\infty(X)$. If γ is a closed geodesic of X, from this equality we infer that

$$\int_\gamma h = \int_\gamma f.$$

Our hypothesis on h therefore implies that the Radon transform $\hat{f}$ of f vanishes. Since the Radon transform is injective on $\mathbb{RP}^2$, the function f vanishes, and so $h = \mathcal{L}_\xi g$.

3. The complex quadric

Let $n \geq 4$. We henceforth suppose that X is the complex quadric Q_n, which is the complex hypersurface of complex projective space $\mathbb{CP}^{n+1}$ defined by the homogeneous equation

$$\zeta_0^2 + \zeta_1^2 + \cdots + \zeta_{n+1}^2 = 0,$$

where $\zeta = (\zeta_0, \zeta_1, \ldots, \zeta_{n+1})$ is the standard complex coordinate system of $\mathbb{C}^{n+2}$. Let g be the Kähler metric on X induced by the Fubini-Study metric on $\mathbb{CP}^{n+1}$ of constant holomorphic curvature 4.

We may identify X with the irreducible Hermitian symmetric space

$$SO(n+2)/SO(2) \times SO(n).$$

Hence it is also an Einstein manifold.

We discuss the various ingredients which enter into the proof of the following result:

THEOREM 3.1. *The complex quadric* Q_n *is infinitesimally rigid for* $n \geq 4$.

First, we require explicit descriptions of the infinitesimally rigid totally geodesic submanifolds of Q_n. These include flat 2-tori and complex projective spaces $\mathbb{CP}^2$. To exploit the infinitesimal rigidity of these spaces, we consider the family $\mathcal{F}$ of closed totally geodesic surfaces of X isometric either to $\mathbb{CP}^1$ with its metric of constant curvature 4, or to $\mathbb{RP}^2$ with its metric of constant curvature 1, or to a flat 2-torus. According to [4], an element of $\mathcal{F}$ isometric to $\mathbb{CP}^1$ or to $\mathbb{RP}^2$ is contained in a totally geodesic submanifold of X isometric to $\mathbb{CP}^2$, with its metric of constant holomorphic curvature 4. We introduce the sub-bundle N of G consisting of those elements of G, which vanish when restricted to the surfaces of $\mathcal{F}$ (see [7, §5] and [8, §5]). The infinitesimal orbit of the curvature $\widetilde{G}$ is a sub-bundle of N, and we identify $N/\widetilde{G}$ with a sub-bundle of $G/\widetilde{G}$. If $\beta : G/\widetilde{G} \to G/N$ is the canonical projection, we consider the differential operator

$$D_1' = \beta D_1 : S^2 T^* \to \mathcal{G}/\mathcal{N}.$$

From the infinitesimal rigidity of $\mathbb{CP}^2$ and of flat 2-tori (due to Michel [13], [14] and Tsukamoto [17]), using the naturality of the operator D_1, we obtain the following:

PROPOSITION 3.1. *If h is an element of $C^\infty(S^2 T^*)$ satisfying the zero-energy condition, then $D_1' h = 0$.*

In [7], we prove the following

PROPOSITION 3.2. *If $n \geq 5$ and $h \in C^\infty(S^2 T^*)$ satisfies $D_1' h = 0$, then $D_1 h = 0$.*

If $n \geq 5$ and h is an element of $C^\infty(S^2 T^*)$ satisfying the zero-energy condition, then by Propositions 3.1 and 3.2 we know that $D_1 h = 0$. By the exactness of the sequence (1.9), we infer that $h \in D_0 C^\infty(T)$, and so Q_n is infinitesimally rigid in this case. The proof of Proposition 3.2 depends on a decomposition of N which holds only when $n \geq 5$ and on a construction of an explicit complement of $\widetilde{G}$ in N.

We now introduce decompositions of the space of symmetric 2-forms. The complex structure of Q_n induces decompositions

$$\textstyle\bigwedge^2 T^* = (\bigwedge^2 T^*)^+ \oplus (\bigwedge^2 T^*)^-, \qquad S^2 T^* = (S^2 T^*)^+ \oplus (S^2 T^*)^-,$$

where $(\bigwedge^2 T^*)^+$ and $(S^2 T^*)^+$ are the sub-bundles of Hermitian forms, while $(\bigwedge^2 T^*)^-$ and $(S^2 T^*)^-$ are the sub-bundles of skew-Hermitian forms. We now use the differential geometry of X considered as a complex hypersurface of $\mathbb{CP}^{n+1}$, which has been studied by Smyth [15]. The components of the second fundamental form of X in $\mathbb{CP}^{n+1}$ generate a sub-bundle E of $(S^2 T^*)^-$ of rank 2 and determine an involution of $(S^2 T^*)^+$. We therefore obtain a decomposition

$$(S^2 T^*)^+ = (S^2 T^*)^{++} \oplus (S^2 T^*)^{+-}$$

of $(S^2 T^*)^+$ into the direct sum of the eigenbundles $(S^2 T^*)^{++}$ and $(S^2 T^*)^{+-}$ corresponding to the eigenvalues $+1$ and -1, respectively, of this involution. In [7], we consider natural monomorphisms of vector bundles

$$\psi : \textstyle\bigwedge^2 T^* \to G, \qquad \hat{\tau} : S^2 T^* \to G;$$

the morphism ψ depends only on the Kähler form of X, while $\hat{\tau}$ depends only on g.

In [8], when $n = 4$ we define an involution $*$ of $(S^2 T^*)^{+-}$, which is uniquely determined up to a sign.

If $n \geq 5$, we are able to give an explicit decomposition of N; in fact, in [7] and [8] we show that

$$(3.1) \qquad N = \tilde{G} \oplus \psi((\wedge^2 T^*)^-) \oplus \hat{\tau}(E).$$

By means of the compatibility condition for the operator D_1, we prove the following:

PROPOSITION 3.3. *If $n \geq 4$ and $h \in S^2 T^*$ satisfies*

$$D_1 h = \alpha(\hat{\tau}(v) + \psi(\beta)),$$

where $v \in \mathcal{E}$ and $\beta \in (\wedge^2 T^*)^-$, then $v = 0$ and $\nabla \beta = 0$.

Since the only harmonic 2-forms on X are the multiples of the Kähler form, Proposition 3.2 follows immediately from the decomposition (3.1) and Proposition 3.3.

When $n = 4$, the decomposition (3.1) fails to hold; the bundle N has several additional components and Proposition 3.2 can no longer be used to prove infinitesimal rigidity. We now outline a new approach to the study of infinitesimal rigidity which we apply to the quadric Q_4. It is partially inspired by Michel's results for $\mathbb{RP}^2$ described in §2.

To prove the infinitesimal rigidity of $X = Q_4$, it suffices to show that any $h \in C^\infty(S^2 T^*)$ satisfying the zero-energy condition and which is orthogonal to $D_0 C^\infty(T)$ vanishes. According to the decomposition (1.1) and Proposition 3.1, we need only show that, for $h \in C^\infty(S^2 T^*)$, the two conditions

$$(3.2) \qquad D_1' h = 0, \qquad \operatorname{div} h = 0$$

imply that h vanishes, or in other words that the operator

$$\operatorname{div} \oplus D_1' : C^\infty(S^2 T^*) \to C^\infty(T^*) \oplus C^\infty(G/N)$$

is injective. We now present the proof of the infinitesimal rigidity of Q_4 as given in [8] and [9]. We shall not require the exactness of the sequence (1.9) as we did in dimension ≥ 5.

Although we are no longer able to decompose N explicitly when $n = 4$, we can still characterize the sub-bundle $\operatorname{Tr} N$ of $S^2 T^*$. We henceforth assume that $X = Q_4$. In [8], we verify the following facts which are considerably simpler to prove than the decomposition (3.1).

PROPOSITION 3.4. *We have the equalities*

$$(3.3) \qquad \operatorname{Tr} N \subset E \oplus (S^2 T^*)^{+-}, \qquad \operatorname{Tr} \cdot \operatorname{Tr} N = \{0\}.$$

We recall that Q_4 is an Einstein manifold with $\operatorname{Ric} = 8g$. We begin by successively proving that an element h of $C^\infty(S^2 T^*)$ satisfying (3.2) has the following properties:

(i) $\Delta \operatorname{Tr} h = 8 \operatorname{Tr} h$;
(ii) $\Delta h - 16h \in C^\infty(E \oplus (S^2 T^*)^{+-})$;

(iii) $h \in C^\infty(E \oplus (S^2T^*)^{+-})$.

Since the first eigenvalue of the Laplacian is equal to 16, we see that (i) implies that $\operatorname{Tr} h = 0$. Properties (i) and (ii) are provided by the following proposition; they represent a crucial step in our proof of the infinitesimal rigidity of Q_4.

PROPOSITION 3.5. *Let h be an element of $C^\infty(S^2T^*)$. If $\operatorname{div} h = 0$ and $D_1'h = 0$, then $\operatorname{Tr} h = 0$ and*

$$(3.4) \qquad \Delta h - 16h \in C^\infty(E \oplus (S^2T^*)^{+-}).$$

PROOF. Our hypotheses imply that $D_1 h$ is a section of $N/\widetilde{G}$. By Proposition 1.2 and the second relation of (3.3), we have

$$\operatorname{Tr}(\operatorname{Ric}_g' h - 8h) = -\operatorname{Tr} \cdot \operatorname{Tr} D_1 h = 0.$$

According to (1.7), we therefore obtain the equality

$$\Delta \operatorname{Tr} h = 8 \operatorname{Tr} h.$$

By relation (4.4) of [8], the first non-zero eigenvalue λ_1 of the Laplacian Δ acting on $C^\infty(X)$ is 16. Hence we see that $\operatorname{Tr} h = 0$. Now from the inclusion of (3.3), we infer that

$$\operatorname{Tr} D_1 h \in C^\infty(E \oplus (S^2T^*)^{+-}).$$

On the other hand, since $\operatorname{Tr} h = 0$, Proposition 1.2 and the identity (1.6) imply that

$$-\operatorname{Tr} D_1 h = \operatorname{Ric}_g' h - 8h = \tfrac{1}{2}\Delta h - 8h.$$

From these last two relations, we deduce (3.4).

The first assertion of Proposition 3.5 is the analogue for Q_4 of Lemma 2.4. While for $\mathbb{RP}^2$ the conditions $\operatorname{div} h = 0$ and $\operatorname{Tr} h = 0$ imply that $\Delta h - 2\lambda h = 0$, here these same conditions and $D_1'h = 0$ only entail that $\Delta h - 2\lambda h$ is a section of $E \oplus (S^2T^*)^{+-}$. For $\mathbb{RP}^2$, the eigenvalues of the Lichnerowicz Laplacian are $\geq 4\lambda$, and so from the relation $\Delta h - 2\lambda h = 0$ we deduce that h vanishes. To obtain the same conclusion in the case of Q_4, we are obliged to use the harmonic analysis on the homogeneous space Q_4 of $SO(6)$.

Let Γ denote the set of irreducible $SO(6)$-modules. If F is a homogeneous unitary Hermitian vector bundle over X, we denote by $C_\gamma^\infty(F)$ the isotypic component of the $SO(6)$-module $C^\infty(F)$ corresponding to $\gamma \in \Gamma$. The vector bundles $F_1 = S^2T_\mathbb{C}^*$ and $F_2 = T_\mathbb{C}^* \oplus (G/N)_\mathbb{C}$, endowed with the Hermitian scalar products induced by the metric g, are homogeneous and unitary; moreover the differential operator

$$Q = \operatorname{div} \oplus D_1' : \mathcal{F}_1 \to \mathcal{F}_2$$

is homogeneous. For $\gamma \in \Gamma$, we consider the $SO(6)$-submodule

$$M_\gamma = C_\gamma^\infty(S^2T_\mathbb{C}^*) \cap \operatorname{Ker} Q$$

of the isotypic component $C_\gamma^\infty(S^2T_\mathbb{C}^*)$.

Using the branching law of [18], in [8] we determine the multiplicities of the $SO(6)$-modules $C_\gamma^\infty(F)$, where F is an invariant complex sub-bundle of $S^2 T_{\mathbb{C}}^*$ and $\gamma \in \Gamma$. We use our knowledge of these multiplicities to prove the following:

PROPOSITION 3.6. *For $\gamma \in \Gamma$, we have*

$$M_\gamma \subset C_\gamma^\infty(E_{\mathbb{C}} \oplus (S^2 T^*)_{\mathbb{C}}^{+-}).$$

PROOF. According to [11], Δ is a multiple of the Casimir operator of the $SO(6)$-module $C^\infty(S^2 T_{\mathbb{C}}^*)$. Therefore, when $C_\gamma^\infty(S^2 T_{\mathbb{C}}^*)$ is non-zero, it is an eigenspace of Δ, whose eigenvalue c_γ depends only on γ. From Proposition 3.5, we deduce that an element h of M_γ satisfies

$$(c_\gamma - 16)h \in C^\infty(E_{\mathbb{C}} \oplus (S^2 T^*)_{\mathbb{C}}^{+-}).$$

Thus if $c_\gamma \neq 16$, we obtain the desired result. In [8], we verify that there is a unique $\gamma_0 \in \Gamma$ for which $C_{\gamma_0}^\infty(S^2 T_{\mathbb{C}}^*)$ is non-zero and $c_{\gamma_0} = 16$. From the computation of the multiplicity of $C_{\gamma_0}^\infty(F)$, where F is an invariant complex sub-bundle of $S^2 T_{\mathbb{C}}^*$, we infer that an element h of $C_{\gamma_0}^\infty(S^2 T_{\mathbb{C}}^*)$ can be written in the form

$$h = fg + h',$$

where f is a complex-valued function on X and h' is an element of $C_{\gamma_0}^\infty(E_{\mathbb{C}} \oplus (S^2 T^*)_{\mathbb{C}}^{+-})$. If h belongs to M_{γ_0}, by Proposition 3.5 we know that $\mathrm{Tr}\, h = 0$, and therefore h belongs to $C_{\gamma_0}^\infty(E_{\mathbb{C}} \oplus (S^2 T^*)_{\mathbb{C}}^{+-})$.

Here the fact that Δ is a multiple of the Casimir operator of the $SO(6)$-module $C^\infty(S^2 T_{\mathbb{C}}^*)$ in some sense replaces Lemma 2.1, which we needed in the case of the real projective plane.

By Proposition 2.2 of [6], $\bigoplus_{\gamma \in \Gamma} M_\gamma$ is a dense subspace of $\mathrm{Ker}\, Q$. Therefore from Proposition 3.6 we deduce step (iii), namely:

PROPOSITION 3.7. *Let h be an element of $C^\infty(S^2 T^*)$. If $\mathrm{div}\, h = 0$ and $D_1' h = 0$, then*

$$h \in C^\infty(E \oplus (S^2 T^*)^{+-}).$$

We next examine the action of the operator div on $C^\infty(E \oplus (S^2 T^*)^{+-})$. We denote by π_{+-} the orthogonal projection of $S^2 T^*$ onto its sub-bundle $(S^2 T^*)^{+-}$.

PROPOSITION 3.8. *If h is an element of $C^\infty(E \oplus (S^2 T^*)^{+-})$ satisfying*

$$\mathrm{div}\, h = 0,$$

then there exists a function $f \in C^\infty(X)$ such that

$$h = *\pi_{+-} \mathrm{Hess}\, f.$$

In particular, Proposition 3.8 tells us that an element h of $C^\infty(E \oplus (S^2 T^*)^{+-})$ satisfying $\mathrm{div}\, h = 0$ is necessarily a section of $(S^2 T^*)^{+-}$. We consider the fourth-order differential operator

$$D_2 = D_1' * \pi_{+-} \mathrm{Hess}$$

acting on functions.

PROPOSITION 3.9. *A function $f \in C^\infty(X)$ solution of $D_2 f = 0$ satisfies $\pi_{+-}\mathrm{Hess}\, f = 0$.*

Propositions 3.7, 3.8 and 3.9, taken together, imply

PROPOSITION 3.10. *An element h of $C^\infty(S^2 T^*)$ satisfying $\mathrm{div}\, h = 0$ and $D_1' h = 0$ vanishes.*

As we have seen above, the infinitesimal rigidity of Q_4 follows directly from this proposition.

To prove Proposition 3.8, we suppose that h is a vector of highest weight of an isotypic component of $C^\infty(E_{\mathbb{C}} \oplus (S^2 T^*)_{\mathbb{C}}^{+-})$. Since the multiplicity of $C_\gamma^\infty(E_{\mathbb{C}} \oplus (S^2 T^*)_{\mathbb{C}}^{+-})$ is ≤ 4 for $\gamma \in \Gamma$, using the harmonic analysis on Q_4 as described by Strichartz [16], we are able to write h explicitly in terms of Hessians of certain eigenfunctions of the Laplacian. The condition $\mathrm{div}\, h = 0$ then implies that the multiplicity of

$$C_\gamma^\infty(E_{\mathbb{C}} \oplus (S^2 T^*)_{\mathbb{C}}^{+-}) \cap \mathrm{Ker}\, \mathrm{div}$$

is ≤ 1. To verify Proposition 3.9, we show that the homogeneous differential operator D_2 does not vanish on those eigenfunctions f of the Laplacian of highest weight for which $\pi_{+-}\mathrm{Hess}\, f \neq 0$. We study the action of the operators div and D_2 on these specific elements of $C^\infty(S^2 T_{\mathbb{C}}^*)$ by means of the formalism of Kähler geometry on the complex quadric. Auspicious choices for the frames of the tangent bundle lead to particularly simple formulas.

If we were to show directly, by representation theory, that $M_\gamma = 0$ for $\gamma \in \Gamma$, without the crucial step given by Proposition 3.7, the requisite computations would be considerably more complicated. To give an idea of the degree of difficulty, we note that the multiplicity of the modules $C_\gamma^\infty(S^2 T_{\mathbb{C}}^*)$ attains a maximum of 14. Our reduction scheme has two other advantages. It totally eliminates certain series of representations of $SO(6)$. Also, most of the required calculations involve the operator div and not the more intricate operator D_1'. This illustrates the extent to which Proposition 3.7 simplifies the original problem.

REFERENCES

1. A. Besse, *Manifolds all of whose geodesics are closed*, Ergeb. Math. Grenzgeb., Bd. 93, Springer-Verlag, Berlin, Heidelberg, New York, 1978.
2. A. Besse, *Einstein manifolds*, Ergeb. Math. Grenzgeb., 3 Folge, Bd. 10, Springer-Verlag, Berlin, Heidelberg, New York, 1987.
3. M. Berger and D. Ebin, *Some decompositions of the space of symmetric tensors on a Riemannian manifold*, J. Differential Geom. **3** (1969), 379–392.
4. Y. Dieng, *Quelques résultats de rigidité infinitésimale pour les quadriques complexes*, C. R. Acad. Sci. Paris Sér. I Math. **304** (1987), 393–396.
5. J. Gasqui and H. Goldschmidt, *Déformations infinitésimales des espaces riemanniens localement symétriques*. I, Adv. in Math. **48** (1983), 205–285.
6. ———, *Déformations infinitésimales des espaces riemanniens localement symétriques*. II. *La conjecture infinitésimale de Blaschke pour les espaces projectifs complexes*, Ann. Inst. Fourier (Grenoble) **34** (1984), no. 2, 191–226.

7. _____, *Rigidité infinitésimale des espaces projectifs et des quadriques complexes*, J. Reine Angew. Math. **396** (1989), 87–121.

8. _____, *On the geometry of the complex quadric*, Hokkaido Math. J. **20** (1991), 279–312.

9. _____, *The infinitesimal rigidity of the complex quadric of dimension four* (to appear).

10. H. Goldschmidt, *The Radon transform for symmetric forms on real projective spaces*, Contemp. Math. **113** (1990), 81–96.

11. N. Koiso, *Rigidity and stability of Einstein metrics – The case of compact symmetric spaces*, Osaka J. Math. **17** (1980), 51–73.

12. A. Lichnerowicz, *Propagateurs et commutateurs en relativité générale*, Inst. Hautes Études Sci. Publ. Math. **10** (1961), 52 pp.

13. R. Michel, *Problèmes d'analyse géométrique liés à la conjecture de Blaschke*, Bull. Soc. Math. France **101** (1973), 17–69.

14. _____, *Un problème d'exactitude concernant les tenseurs symétriques et les géodésiques*, C. R. Acad. Sci. Paris Sér. A **284** (1977), 183–186; *Tenseurs symétriques et géodésiques*, C. R. Acad. Sci. Paris Sér. A **284** (1977), 1065–1068.

15. B. Smyth, *Differential geometry of complex hypersurfaces*, Ann. of Math. **85** (1967), 246–266.

16. R. Strichartz, *The explicit Fourier decomposition of $L^2(SO(n)/SO(n-m))$*, Can. J. Math. **27** (1975), 294–310.

17. C. Tsukamoto, *Infinitesimal Blaschke conjectures on projective spaces*, Ann. Sci. École Norm. Sup. (4) **14** (1981), 339–356.

18. _____, *Spectra of Laplace-Beltrami operators on $SO(n+2)/SO(2) \times SO(n)$ and $Sp(n+1)/Sp(1) \times Sp(n)$*, Osaka J. Math. **18** (1981), 407–426.

DEPARTMENT OF MATHEMATICS, COLUMBIA UNIVERSITY, NEW YORK, NEW YORK 10027

E-mail address: hg@shire.math.columbia.edu

Contemporary Mathematics
Volume **140**, 1992

Microlocal Analysis of
the Two-Plane Transform

ALLAN GREENLEAF AND GUNTHER UHLMANN

Introduction

Let $M_{2,n}$ be the grassmannian of affine 2-planes in R^n . Define, initially for $f \in C_0^\infty(R^n)$, the two-plane transform [H]

$$(R_{2,n}f)(\Pi) = \int_\Pi f(y)d\sigma(y), \quad \Pi \in M_{2,n},$$

where $d\sigma(y)$ is normalized Lebesgue measure on Π . $R_{2,n}$ extends to $\mathcal{E}'(R^n)$ by duality:

$$\langle R_{2,n}u, g \rangle = \langle u, R_{2,n}^*g \rangle, \quad u \in \mathcal{E}'(R^n), g \in C^\infty(M_{2,n}),$$

where $R_{2,n}^*$ is the adjoint operator

$$(R_{2,n}^*g)(x) = \int_{x+G_{2,n}} g(\Pi)d\Pi.$$

The range of $R_{2,n}$ can be characterized by a system of second order partial differential equations [GGS] or by a single fourth order equation [Gn] . Since $R_{2,n}$ is an overdetermined operator, it is of interest to find sets of uniqueness. That is, if $\mathcal{C} \subset M_{2,n}$ is a subset, and we define the *restricted* two-plane transform

$$R_{\mathcal{C}}f = R_{2,n}f|_{\mathcal{C}},$$

one wishes to determine for which $\mathcal{C}$ the operator $R_{\mathcal{C}}$ is injective on various function spaces and find inversion formulas. A "universal" local inversion formula was found in [GG], and extensive further work can be found in [Go1,2,3], but even in the complex setting, the problem of determining which submanifolds $\mathcal{C} \subset M_{2,n}$ are admissible in the sense of admitting a local inversion formula seems

1991 *Mathematics Subject Classification.* Primary 44A05; Secondary 47G05.

Both authors partially supported by the NSF. The first author also supported by an Alfred P. Sloan Research Fellowship.

This paper is in final form and no version of it will be submitted for publication elsewhere.

to remain open. Over the reals, one must deal with nonlocal inversion formulas, which were first investigated in [GGi]; more recent results are in [GG2]. We study this problem in the microlocal setting, continuing earlier work on the x-ray (one-plane) transform [GrU1,3,4]. This is intended as an expository article which emphasizes the microlocal geometry induced by the admissibility criteria. The corresponding microlocal analysis will be found in [GrU5].

Admissibility

As is well-known, there is a natural embedding

$$\hat{}: M_{2,n} \hookrightarrow G_{3,n+1}$$

of $M_{2,n}$ as an open subset of the grassmannian of three-dimensional subspaces of R^{n+1} . For $\Pi \in M_{2,n}$ there is thus a natural identification

$$T_{\Pi}^* M_{2,n} \simeq T_{\hat{\Pi}}^* G_{3,n+1} \simeq \hat{\Pi} \otimes (R^{n+1}/\hat{\Pi})^*.$$

Let $\mathcal{C} \subset M_{2,n}$ be a submanifold of dimension $n-2+m$, so that $m = 2$ corresponds to a complex of two-planes and $m = 2n - 5$ to a hypersurface in $M_{2,n}$. The conormal space to $\mathcal{C}$ at a point (i.e., two-plane) Π ,

$$N_{\Pi}^* \mathcal{C} \subset T_{\hat{\Pi}}^* M_{2,n} \simeq \hat{\Pi} \otimes (R^{n+1}/\hat{\Pi})^*,$$

is identified with a $3n-6-(n-2+m) = (2n-4-m)$-dimensional space of linear mappings from $R^{n+1}/\hat{\Pi}$ to $\hat{\Pi}$. For each $\xi \in \hat{\Pi}\backslash 0$, there is a natural evaluation mapping

$$e_{\xi} : N_{\Pi}^* \mathcal{C} \to R^{n+1}/\hat{\Pi}.$$

Our first version of admissibility is a simple extension of Maius' criterion for complex lines [M]. Unfortunately, it only applies when the codimension of $\mathcal{C}$ is relatively small.

DEFINITION 1. $\mathcal{C}$ is admissible in the sense of Maius if for generic $\Pi \in \mathcal{C}$, (a) the mapping e_{ξ} is injective for generic $\xi \in \hat{\Pi}\backslash 0$; and (b) the image $e_{\xi}(N_{\Pi}^* \mathcal{C}) \subset R^{n+1}/\hat{\Pi}$ is independent of ξ for generic $\xi \in \hat{\Pi}\backslash 0$.

Clearly, for (a) to hold, we must have $2n - 4 - m \leq n - 2$,i.e., $m \geq n - 2$, and for (b) to be interesting we must have $m > n - 2$, so that $dim(\mathcal{C}) > 2n - 4$.

The second notion of admissibility, which is a generalization of Gelfand's cone condition for the x-ray transform, is applicable when $n \geq 5$, and is equivalent with the first when they both apply. For generic $x \in R^n$, $\mathcal{C}_x = \{\Pi \in \mathcal{C} : x \in \Pi\}$ is an m-dimensional submanifold of $\mathcal{C}$. Thus,

$$\Gamma_{\Pi} = \bigcup_{x \in \Pi} \mathcal{C}_x$$

is the union of a 2-parameter family of m-surfaces in $\mathcal{C}$; in general this will have a conical singularity at the vertex Π .

DEFINITION 2. *C is admissible in the sense of Gelfand if for generic $\Pi \in C$,
Γ_Π is (C^1)-smooth at Π, i.e., if*

$$\bigcup_{x \in \Pi} T_\Pi C_x$$

is contained in an $(m+2)$-plane in $T_\Pi C$.

In addition to admissibility, a curvature assumption is required to allow the application of the microlocal machinery described below. This is a straightforward extension of the well-curvedness condition in [GrU4].

The main result can now be stated.

THEOREM. *Let $C^{n-2+m} \subset M_{2,n}$ be well-curved and admissible in the sense of either Maius or Gelfand. Then, if either $m = 2$ (so that C is a complex) or the curvature form A defined below is nonsingular, R_C has a relative left parametrix T,*

$$T R_C = I - E \quad on \quad \mathcal{E}'_K(R^n),$$

where E is a sum of Fourier integral operators.

Here, we use Hörmander's notation $\mathcal{E}'_K(R^n)$ for the distributions with compact support and wavefront set contained in a closed set K; E and K are described below.

COROLLARY ("1/4-ESTIMATE"). *$R_C : H^s_{comp}(R^n) \to H^{s+s_0}_{loc}(C)$, where*

$$s_0 = \begin{cases} 1/2, & m = 2 \\ 3/4, & m = 3 \\ 1 - \epsilon, & m = 4, \quad any \quad \epsilon > 0 \\ 1, & m \geq 5. \end{cases}$$

By general considerations [Gu,GuS], R_C is a Fourier integral operator of order $-(m+2)/4$ associated with the canonical relation

$$C = N^* Z'_C \subset T^*C \times T^*R^n,$$

where

$$Z_C = \{(\Pi, x) \in C \times R^n : x \in C\}$$

is the point-plane incidence relation. One needs to understand the structure of the canonical relation C, in terms of the singular structure of the projections $\pi : C \to T^*R^n$ and $\rho : C \to T^*C$. Consider the case when C is a complex, so that C, T^*R^n and T^*C all have dimension $2n$. Insight into the structure of π and ρ in this situation can be gained by considering the following class of models. Let

$$A \in R^2 \otimes S^2 R^{2^*};$$

we think of A both as a trilinear form, symmetric in the last two arguments, and as a linear space of quadratic forms. Writing elements of R^n as $x = (x', x'') \in R^{n-2} \times R^2$, consider the phase function

$$\phi(x, y, \theta') = (x' - y') \cdot \theta' + A(y'', x'', x'')\theta_1, \quad |\theta_1| \geq c|\theta|,$$

on $R^n \times R^n \times (R^{n-2}\backslash 0)$. This arises, for example, in the Fourier integral representation of R_C for the admissible complex $C \subset M_{2,n}$ with Z_C given by (using coordinates y on R^n and x on C)

$$Z_C = \{(x, y) : x_1 - y_1 + A(y'', x'', x'') = 0, x_j - y_j = 0, 2 \leq j \leq n - 2\}.$$

The determinantal form on $S^2 R^{2^*}$ (with respect to some volume form on R^2) is a quadratic form of signature $(1, 2)$. If A is in general position, which we will assume, then, considered as a 2-dimensional subspace of $S^2 R^{2^*}$, away from the origin it intersects the determinantal variety transversally. That is, A is either spacelike or timelike. If A is spacelike, then one can prove that C is a $(2, 2)$-fibered folding canonical relation in the sense of [GrU4]: a parameterization of C similar to that used in the x-ray transform case shows that the projection $\pi : C \to T^* R^n$ is a submersion with clean folds of multiplicity 2 and $\rho : C \to T^*C$ is a $(4, 2, 2)$-blowdown. The image $\pi(L)$ of the fold surface $L \subset C$ is a codimension two involutive submanifold of $T^* R^n$. The construction of the nonlocal inversion formula then follows from the composition calculus of [GrU4]: $R^*_C R_C$ belongs to the Guillemin-Melrose-Uhlmann class $I^{p,l}(\Delta_{T^* R^n}, \Lambda_{\pi(L)})$ for appropriate p, l, where $\Delta_{T^* R^n}$ is the diagonal and $\Lambda_{\pi(L)}$ is the flowout of $\pi(L)$. From the known composition calculus for the $I^{p,l}$ classes [AU], one constructs a relative left parametrix , with error $E \in I^{-1}(\Lambda_{\pi(L)})$ and $K = \overline{\pi(C)}$. From the estimates for the $I^{p,l}$'s [GrU2], one obtains the Corollary in this case.

Thus, any new analysis concerns the case of A timelike. An example is given by

$$A = \frac{1}{2}(e_1 \otimes f_1^* \otimes f_1^* + e_2 \otimes f_2^* \otimes f_2^*),$$

where e_1, e_2 is a basis for R^2 and f_1^*, f_2^* is a basis for R^{2^*} , so that

$$\phi(x, y, \theta') = (x' - y') \cdot \theta' + \frac{1}{2}(y_{n-1}x_{n-1}^2 + y_n x_n^2)\theta_1,$$

which parameterizes the canonical relation

$$C_\phi = \{(x', x'', \xi', y_{n-1}x_{n-1}\xi_1, y_n x_n\xi_1; x_1 + \frac{1}{2}A(y'', x'', x''), x_2, ..., x_{n-2}, y'',$$

$$\xi', -\frac{1}{2}x_{n-1}^2\xi_1, -\frac{1}{2}x_n^2\xi_1) : \quad x \in R^n, \xi' \in R^{n-2}\backslash 0, y'' \in R^2\}$$

$$= \{((y_1 - \frac{1}{2}A(y'', x'', x''), y_2, ..., y_{n-2}, x'', \eta', y_{n-1}x_{n-1}\eta_1, y_n x_n\eta_1;$$

$$y', y'', \eta', -\frac{1}{2}x_{n-1}^2\eta_1, -\frac{1}{2}x_n^2\eta_1) : x'' \in R^2, y \in R^n, \eta' \in R^{n-2}\backslash 0\}.$$

From the first parameterization, we see that the projection ρ onto the first factor is a product of two $(2,1,1)$-blowdowns (i.e., each of the same singularity class as polar coordinates in R^2) and from the second that the projection π onto the second factor is a product of Whitney folds. The singular locus L, where C fails to be a local canonical graph, is

$$L = \{x_{n-1} = 0\} \cup \{x_n = 0\} = L_1 \cup L_2,$$

the union of two hypersurfaces, meeting transversally. The images $\pi(L_1)$ and $\pi(L_2)$ are hypersurfaces in $T^* R^n$, meeting transversally in a codimension two involutive submanifold. In constructing nonlocal inversion formulas, we again consider the composition $R_C^* R_C$. By the Hörmander-Sato Lemma,

$$WF(R_C^* R_C)' \subset C^t \circ C.$$

In the above example, on easily finds that $C^t \circ C$ lies in the union of four intersecting canonical relations in $T^* R^n \times T^* R^n$,

$$C^t \circ C \subset \Lambda_0 \cup \Lambda_1 \cup \Lambda_2 \cup \Lambda_3,$$

where

$$\Lambda_0 = \Delta_{T^* R^n} = \quad the \quad diagonal;$$
$$\Lambda_1 = \quad the \quad flowout \quad of \quad \pi(L_1);$$
$$\Lambda_2 = \quad the \quad flowout \quad of \quad \pi(L_2); \quad and$$
$$\Lambda_3 = \quad the \quad joint \quad flowout \quad of \quad \pi(L_1) \cap \pi(L_2).$$

The 4-tuple $\{\Lambda_j\}_{j=0}^3$ forms an intersecting system of lagrangians in the sense of [MeU], where it was shown that parametrices for some operators with involutive double characteristics belong to classes of operators with wavefront set lying in such intersecting systems. Extending the distribution classes in [MeU] to have more general symbols, we associate to each intersecting system a class

$$I^{p,\{l_j\}_{j=1}^3}(\Lambda_0, \Lambda_1, \Lambda_2, \Lambda_3).$$

When the Λ's are the diagonal and flowouts as above, we establish a composition calculus using a variant of the iterated regularity argument in [GrU3]. From the calculus, sharp L^2 estimates and the construction of relative left parametrices easily follow. In this case, the operator E is a sum of Fourier integral operators associated with Λ_1 , Λ_2 and Λ_3 , of orders -1/2 , -1/2 and -1, respectively.

To go from the above model to the general situation needed in the Theorem, one needs to formulate singularity classes that π and ρ belong to, and show that the projections from $N^* Z'_C$ belong to these classes. One then defines the general class of canonical relations having this structure and shows that if S and T are Fourier integral operators associated with such a canonical relation, then

$$S^* T \in I^{p,\{l_j\}_{j=1}^3}(\Lambda_0, \Lambda_1, \Lambda_2, \Lambda_3)$$

for appropriate p and $\{l_j\}_{j=1}^3$, reducing the problem to the calculus for $I^{p,\{l_j\}_{j=1}^3}(\Lambda_0, \Lambda_1, \Lambda_2, \Lambda_3)$.

To deal with admissible $\mathcal{C} \subset M_{2,n}$ of dimension $n-2+m$, one would have to extend the above analysis to include the situation where $A \in R^2 \otimes S^2 R^{m^*}$ is in general position. When A, which is really the Hessian of the projection π at the corresponding points of C, is *nonsingular* in the sense of [GrU4], that is, only consists of rank m elements (other than 0), then C is a $(2, m)$-fibered folding canonical relation and the results of [GrU4] still apply. However, for general A, this leads to the consideration of intersecting systems of (up to) $m+2$ lagrangians. There is unfortunately no normal form for such intersecting systems, making the proof of a composition calculus more difficult. We hope to return to this in the future.

REFERENCES

[AU]. J. Antoniano and G. Uhlmann, *A functional calculus for a class of pseudodifferential operators with singular symbols*, Proc.Symp.Pure Math. **43** (l985), 5-16.

[GGi]. I.M. Gelfand and G.S. Gindikin, *Nonlocal inversion formulas in real integral geometry*, Func.An. Appl. **11** (1977), 173-179.

[GG]. I.M. Gelfand and M.I. Graev, *Line complexes in the space C^n*, Fun.An.Appl. **2** (1968), 219-229.

[GG2]. I.M. Gelfand and M.I. Graev, *Crofton's Function and Inversion Formulas in Real Integral Geometry*, Fun.An.Appl. **25** (1991), 1-5.

[GGS]. I.M. Gelfand, M.I. Graev and Z.Ya. Shapiro, *Integral geometry on k-dimensional planes*, Fun.An.Appl. **1** (1967), 14-27.

[Go1]. A.B. Goncharov, *Integral geometry on families of surfaces in the space*, Jour. Geo. Phy. **5** (1988), 571-593.

[Go2]. A.B. Goncharov, *Integral geometry on families of k–dimensional submanifolds*, Fun. An. Appl. **23** (1989), 178-189.

[Go3]. A.B. Goncharov, *Integral Geometry and manifolds of minimal degree in CP^n*, Fun. An. Appl. **24** (1990), 4-17.

[Gn]. F. Gonzalez, *Invariant differential operators and the range of the Radon d–plane transform*, Math. Annalen 287 (1990), 627-635.

[GrU1]. A. Greenleaf and G. Uhlmann, *Nonlocal inversion formulas for the X-ray transform*, Duke Math.Jour. **58** (1989), 205-240.

[GrU2]. A. Greenleaf and G. Uhlmann, *Estimates for singular Radon transforms and pseudo-differential operators with singular symbols*, Jour.Func.An. **89** (1990), 202-232.

[GrU3]. A. Greenleaf and G. Uhlmann, *Composition of some singular Fourier integral operators and estimates for restricted X-ray transforms*, Ann. Inst. Four. **40** (1990), 443-466.

[GrU4]. A. Greenleaf and G.Uhlmann, *Composition of some singular Fourier integral operators and estimates for restricted X-ray transform,II*, Duke Math. Jour. **64** (1991), 415-444.

[GrU5]. A. Greenleaf and G.Uhlmann, *Quadruple lagrangian intersection and restricted two-plane transforms*, in preparation.

[Gu]. V. Guillemin, *On some results of Gelfand in integral geometry*, Proc. Symp. Pure Math. **43** (1985), 149-155.

[GuS]. V. Guillemin and S. Sternberg, *Geometric Asymptotics*, Amer. Math.Soc., Providence, 1977.

[GU]. V. Guillemin and G. Uhlmann, *Oscillatory integrals with singular symbols*, Duke Math. Jour. **48** (1981), 251-267.

[H]. S. Helgason, *The Radon Transform,Birkhaüser*, Boston, 1980.

[M]. K. Maius, *The structure of admissible line complexes in C^n*, Trans.Moscow Math. Soc.(translated) **39** (1981), 195-226.

[MeU]. R. Melrose and G. Uhlmann, *Lagrangian intersection and the Cauchy problem*, Comm. Pure Appl. Math. **32** (1979), 482-512.

UNIVERSITY OF ROCHESTER, ROCHESTER NY 14627

UNIVERSITY OF WASHINGTON, SEATTLE WA 98195

Contemporary Mathematics
Volume **140**, 1992

Aspects of Flat Radon transforms

ERIC L. GRINBERG *

ABSTRACT. Starting with X a compact symmetric space we consider the operator of integration along maximal totally geodesic flat submanifolds. By working out some explicit examples we show that either this transform is injective or else there is an obvious kernel arising from symmetry considerations and, passing to a quotient, we obtain injectivity.

§0. Integrals over totally geodesic flat submanifolds

In the early part of this century P. Funk showed that an even function on the sphere S^2 is determined by its integrals over great circles and J. Radon showed that a function on R^2 (suitably integrable) is determined by its integrals over straight lines. These results were motivated by problems in geometry and analysis. They led to vast extensions and generalizations. Here we will consider some integral transforms that are direct analogs of Funk's and, like the great circle transform, have applications to geometry. We will present the analysis for some concrete examples. These provide independent proofs of specializations of some general results that will appear elsewhere in a more abstract setting [Gr, 1991]. The approach presented here has the advantage that it may lead to an inversion formula. We hope that this exposition will serve both to illustrate the main theorem of [Gr, 1991] and to make more accessible the explicit harmonic analysis and geometry of symmetric spaces.

The Radon transform and, more generally, the k-plane transform assigns to a function in R^n its integrals over planes of a fixed dimension. These planes are the *flat* totally geodesic submanifolds of Euclidean space; the hyperplanes

1991 *Mathematics Subject Classification*. Primary 44A12; Secondary 53C55, 43A85.

Key words and phrases. Radon transform, symmetric spaces, flat totally-geodesic submanifolds, group representations.

*Supported by a grant from the National Science Foundation

This paper is in final form and no version of it will be submitted for publication elsewhere

are the maximal such. The integral transforms are interesting for their own sake and prove useful in certain analytic and geometric applications. There are many ways to generalize these transforms, but two of the ways seem to keep the flats in greatest focus. In the first scenario we replace Euclidean space by a Riemannian manifold X and integrate over the flat totally geodesic submanifolds of this new geometry. Alternatively, we can let X be the *variety* of all flat totally geodesic k-dimensional submanifolds of Euclidean space and consider integrals over submanifolds of this family of flats. We will combine these two approaches by adorning the space of flats in Euclidean space with a Riemannian structure, and then consider flats in this new geometry. As with the classic Radon transform, the resulting integral operator is both interesting for its own sake *and* useful in applications.

We will examine only *symmetric* Riemannian manifolds [H, 1978]. These are complete manifolds X which, for each $p, q \in X$, have an isometry Φ which transposes p and q. The *rank* of such a manifold X is the dimension of its maximal totally geodesic flat submanifolds. Thus the sphere is of rank 1 while R^n is of rank n. Before treating the general situation we examine the rank 1 case. While the entire analysis may be carried out starting in the *real* domain, the complex case is far cleaner and provides a convenient model for post-processing of the real case. Although we assume some familiarity with homogeneous spaces such as projective spaces, we will review analysis constructions on these as we need them . See [BGM], [H, 1978, 1984], [S, 1975, 1977, 1981], [Gr, 1983, 1986] for further discussion of these manifolds and their associated functions spaces. V. Guillemin has shown that this integral transform gives an obstruction to the existence of conformal isospectral deformations of compact symmetric space; see [GKK].

§1. The maximal flat totally geodesic transform on Grassmannians

The complex projective space CP^n is the space of all complex lines through the origin in C^{n+1}. A unit vector $\omega \in C^{n+1}$ spans a complex line $[\omega] \in CP^n$. The map $\omega \mapsto [\omega]$ gives rise to the Hopf fibration $S^{2n+1} \longrightarrow CP^n$ and the geometry on the base space can be read off from the geometry of the total space. In particular, a maximal totally geodesic flat submanifold in CP^n is simply a geodesic (CP^n, like S^{2n+1}, is of rank 1). The maximal totally geodesic flat transform in this context is just the *X-ray transform*. The complex projective *line* CP^1 is (isomorphic to) the sphere S^2, and its geodesics are the great circles.

Since odd functions on S^2 have vanishing integrals over all great circles, the flat Radon transform on CP^1 has a kernel. For the rest of the complex projective spaces $\{CP^n, n > 1\}$ the flat Radon transform is injective. This was proved by Helgason [H, 1981] who viewed it as an X-ray transform. His proof simply expresses the maximal totally geodesic (not necessarily flat) transform in terms of the X-ray transform, and then invokes the injectivity of the former transform, a fact proved earlier by Helgason [H, 1965] (see also Grinberg [Gr, 1983]). In this context, the maximal totally geodesic submanifolds are projective hyperplanes in $\mathbf{C}P^n$; since each hyperplane is a union of geodesics , the flat Radon transform determines the maximal totally geodesic transform, and this in turn may be inverted by Helgason's inversion formula. It is not hard to see the injectivity of the X-ray transform on $\mathbf{C}P^n$ directly. Consider the following circle of points on $\mathbf{C}P^n$ expressed in homogeneous coordinates:

$$T = \{\, [e^{\sqrt{-1}\theta}/\sqrt{3}, 0, \dots, 0, 1/\sqrt{3}, e^{\sqrt{-1}\theta}/\sqrt{3}] \mid \theta \in \mathbf{R} \,\}.$$

This is the Hopf-image of a great circle in S^{2n+1} which is transverse to the fibers of the Hopf map. To show that T is a geodesic, we need to exhibit it as the image of a great circle that is *orthogonal* to the fibers. To do this, simply multiply the homogeneous coordinates in the definition of T by the scalar factor $e^{-\sqrt{-1}\theta/2}$. This gives an equivalent presentation of T that comes from a fiber-orthogonal great circle and hence is a geodesic. Consider now the function

$$f_m[z_0, \dots, z_n] \equiv (z_0 \overline{z_n})^m,$$

for m a non-negative integer. If we restrict the homogeneous coordinates $(z_0, \dots, z_n)$ to the sphere S^{2n+1} then $f_m(z)$ is a well defined function (otherwise it only makes sense as a section of a line bundle). Clearly,

$$\int_T f_m(z) \neq 0.$$

Thus the kernel of the X-ray transform does not contain any of the functions f_m. On the other hand, if the X-ray transform does have a kernel, this kernel is a $U(n+1)$-invariant function space (loosely, a rotation-invariant space), and standard arguments using representation theory or spherical harmonics show that this space must contain some f_m (see Dieudonné [D]). We will need an extension of this argument later so we summarize it here as follows:

WORKING PRINCIPLE. *A rotation-invariant subspace of functions on $\mathbf{C}P^n$ is completely determined by the functions of the form f_m that it contains.*

Strictly speaking, this principle applies to L^2-closed subspaces only.

We now pass to the general complex Grassmannian $Gr(k, n)$. This is the space of $k+1$-dimensional vector subspaces of $\mathbf{C}^{n+1}$ (or, equivalently, the space of projective k-planes in $\mathbf{C}P^n$). A $k+1$ plane is spanned by the columns of an $(n+1) \times (k+1)$ matrix of the form

$$\begin{pmatrix} Z_0^0 & \cdots & Z_k^0 \\ \cdot & \cdots & \cdot \\ \cdot & \cdots & \cdot \\ \cdot & \cdots & \cdot \\ Z_0^n & \cdots & Z_k^n \end{pmatrix}.$$

We can take the matrix to be orthogonal (or unitary) by columns. The space of all such matrices is the *Stiefel manifold* $St(k,n) = U(n+1)/U(n-k)$. Unitary change of basis of a Stiefel matrix is realized by a right multiplication by a matrix in $U(k+1)$, so that $Gr(k,n) = St(k,n)/U(k+1)$. It will be convenient to think of functions on the Grassmannian as right-$U(k+1)$ invariant functions on a Stiefel manifold. A Schauder basis of such functions is given by polynomials in the various Hermitian inner products of the rows of a Stiefel matrix. We will need analogs of the distinguished polynomials f_m above. The duality between planes and their orthogonal complements gives the isomorphism $Gr(k,n) = Gr(n-k,n)$. Therefore we assume that $k+1 \leq n-k$. Denote the rows of Stiefel matrix by $Z^0, \dots, Z^n$. For each non-increasing sequence of integers $(m_0, \dots, m_k)$ let

$$
\begin{aligned}
f_{m_0,\dots,m_k} = {}& det \left[(Z^0)(Z^n)^* \right]^{m_0-m_1} \\
& \times det \left[\begin{pmatrix} Z^0 \\ Z^1 \end{pmatrix} \cdot \begin{pmatrix} Z^n \\ Z^{n-1} \end{pmatrix}^* \right]^{m_1-m_2} \\
& \times \cdots \times \\
& \times det \left[\begin{pmatrix} Z^0 \\ \vdots \\ Z^k \end{pmatrix} \cdot \begin{pmatrix} Z^n \\ \vdots \\ Z^{n-k} \end{pmatrix}^* \right]^{m_k-0} .
\end{aligned}
$$

We denote $f_{m_0,\dots,m_k}$ by f_m for brevity and analogy with the rank 1 case. It is clear that each f_m is invariant under right multiplication by a $(k+1) \times (k+1)$ unitary matrix, hence a well defined function on the Grassmannian. It is of course possible to express f_m directly in terms of dot products of rows, but we will not need such an expression here. As with projective spaces, we have the following tool for harmonic analysis:

WORKING PRINCIPLE. *A rotation-invariant subspace of functions on Grassmannian is completely determined by the functions of the form f_m that it contains.*

Here *rotation invariant* means invariant under the isotropy group of a base point. This principle follows from standard results in representation theory. See Strichartz [S] or Grinberg [Gr].

The cases $k+1 < n-k$ and $k+1 = n-k$ are qualitatively different, so we consider the former situation first. Consider now the following variety of Stiefel matrices:

$$\begin{pmatrix}
e^{\sqrt{-1}\theta_0} & 0 & \cdots & 0 & 0 \\
0 & e^{\sqrt{-1}\theta_1} & \cdots & 0 & 0 \\
\vdots & \vdots & \ddots & \vdots & \vdots \\
0 & 0 & \vdots & e^{\sqrt{-1}\theta_{k-1}} & 0 \\
0 & 0 & \vdots & 0 & e^{\sqrt{-1}\theta_k} \\
1 & 1 & \cdots & 1 & 1 \\
0 & 0 & \cdots & 0 & 0 \\
\vdots & \vdots & \ddots & \vdots & \vdots \\
0 & 0 & \cdots & 0 & 0 \\
0 & 0 & \vdots & 0 & e^{\sqrt{-1}\theta_k} \\
0 & 0 & \vdots & e^{\sqrt{-1}\theta_{k-1}} & 0 \\
\vdots & \vdots & \ddots & \vdots & \vdots \\
0 & e^{\sqrt{-1}\theta_1} & \cdots & 0 & 0 \\
e^{\sqrt{-1}\theta_0} & 0 & \cdots & 0 & 0
\end{pmatrix}$$

This is a flat torus in the Grassmannian. In fact, it is the orbit of the plane

$$\begin{pmatrix}
1 & 0 & \cdots & 0 & 0 \\
0 & 1 & \cdots & 0 & 0 \\
\vdots & \vdots & \ddots & \vdots & \vdots \\
0 & 0 & \vdots & 1 & 0 \\
0 & 0 & \vdots & 0 & 1 \\
1 & 1 & \cdots & 1 & 1 \\
0 & 0 & \cdots & 0 & 0 \\
\vdots & \vdots & \ddots & \vdots & \vdots \\
0 & 0 & \cdots & 0 & 0 \\
0 & 0 & \vdots & 0 & 1 \\
0 & 0 & \vdots & 1 & 0 \\
\vdots & \vdots & \ddots & \vdots & \vdots \\
0 & 1 & \cdots & 0 & 0 \\
1 & 0 & \cdots & 0 & 0
\end{pmatrix} \qquad (*)$$

by the torus subgroup

$$\begin{pmatrix}
E(\theta) & 0 & 0 \\
0 & I_{(n-2k)\times(n-2k)} & 0 \\
0 & 0 & E(\tilde{\theta})
\end{pmatrix}$$

where $I_{(n-2k)\times(n-2k)}$ denotes the $(n - 2k - 1) \times (n - 2k - 1)$ identity matrix while $E(\theta)$ and $E(\tilde{\theta})$ denote the $(k + 1) \times (k + 1)$ matrices

$$\begin{pmatrix} e^{\sqrt{-1}\theta_0} & 0 & \cdots & 0 & 0 \\ 0 & e^{\sqrt{-1}\theta_1} & \cdots & 0 & 0 \\ \vdots & \vdots & \ddots & \vdots & \vdots \\ 0 & 0 & \vdots & e^{\sqrt{-1}\theta_{k-1}} & 0 \\ 0 & 0 & \vdots & 0 & e^{\sqrt{-1}\theta_k} \end{pmatrix}$$

and

$$\begin{pmatrix} e^{\sqrt{-1}\theta_k} & 0 & \cdots & 0 & 0 \\ 0 & e^{\sqrt{-1}\theta_{k-1}} & \cdots & 0 & 0 \\ \vdots & \vdots & \ddots & \vdots & \vdots \\ 0 & 0 & \vdots & e^{\sqrt{-1}\theta_1} & 0 \\ 0 & 0 & \vdots & 0 & e^{\sqrt{-1}\theta_0} \end{pmatrix}$$

respectively.

From the general structure theory of symmetric spaces (see Helgason [H, 1978]) it follows that the maximal totally geodesic flats in the Grassmannian are the orbits of maximal tori of the unitary group. While the Stiefel matrix (*) determines the maximal flat that we are going to integrate over, we cannot use it directly to evaluate the function f_m since our original definition requires a *unitary* Stiefel matrix. Of course, we could perform a Gram-Schmidt orthonormalization explicitly, but this is not strictly necessary for our purpose. The submatrices whose determinants appear in the definition of f_m are non-singular and upper (or lower) triangular for our matrix (*). The Gram-Schmidt process preserves the non-singular upper triangular property. Also, the submatrices involved in f_m come in complex conjugate pairs. Therefore, the function f_m is positive throughout the flat torus, and hence its integral is positive. We have shown that

$$\int_T f_m \neq 0,$$

where T is the maximal flat defined by the Stiefel matrices(*). Thus the maximal totally geodesic transform is injective when $k + 1 < n - k$. Of course, the entire construction depends on the row of 1's in the matrix (*). If this row consisted of zeros then the matrix (*) would represent a point and not a torus in the Grassmannian. When the condition $k + 1 < n - k$ holds there is no room for this row and, as we shall see, the results are different.

We illustrate our reasoning by working out the case $Gr(2,6)$. By definition, this is the space of projective 2-planes in projective 6-space, or equivalently, affine 3-planes in affine 7-space. The "raw" Stiefel matrix describing our torus is

$$\begin{pmatrix} e^{\sqrt{-1}\theta_1} & 0 & 0 \\ 0 & e^{\sqrt{-1}\theta_2} & 0 \\ 0 & 0 & e^{\sqrt{-1}\theta_3} \\ 1 & 1 & 1 \\ 0 & 0 & e^{\sqrt{-1}\theta_3} \\ 0 & e^{\sqrt{-1}\theta_2} & 0 \\ e^{\sqrt{-1}\theta_1} & 0 & 0 \end{pmatrix}.$$

After Gram-Schmidt orthonormalization this becomes

$$\begin{pmatrix} \dfrac{e^{\sqrt{-1}\theta_1}}{\sqrt{3}} & -\left(\dfrac{\sqrt{3}e^{\sqrt{-1}\theta_1}}{6\sqrt{2}}\right) & -\left(\dfrac{\sqrt{2}e^{\sqrt{-1}\theta_1}}{4\sqrt{5}}\right) \\[2mm] 0 & \dfrac{\sqrt{3}e^{\sqrt{-1}\theta_2}}{2\sqrt{2}} & -\left(\dfrac{\sqrt{2}e^{\sqrt{-1}\theta_2}}{4\sqrt{5}}\right) \\[2mm] 0 & 0 & \dfrac{\sqrt{2}e^{\sqrt{-1}\theta_3}}{\sqrt{5}} \\[2mm] \dfrac{1}{\sqrt{3}} & \dfrac{\sqrt{3}}{3\sqrt{2}} & \dfrac{\sqrt{2}}{2\sqrt{5}} \\[2mm] 0 & 0 & \dfrac{\sqrt{2}e^{\sqrt{-1}\theta_3}}{\sqrt{5}} \\[2mm] 0 & \dfrac{\sqrt{3}e^{\sqrt{-1}\theta_2}}{2\sqrt{2}} & -\left(\dfrac{\sqrt{2}e^{\sqrt{-1}\theta_2}}{4\sqrt{5}}\right) \\[2mm] \dfrac{e^{\sqrt{-1}\theta_1}}{\sqrt{3}} & -\left(\dfrac{\sqrt{3}e^{\sqrt{-1}\theta_1}}{6\sqrt{2}}\right) & -\left(\dfrac{\sqrt{2}e^{\sqrt{-1}\theta_1}}{4\sqrt{5}}\right) \end{pmatrix}.$$

The triangular property promised in the general discussion is evident here. The determinants occurring in the definition of the function f_m are the principal minors of the matrix below:

$$\begin{pmatrix} \dfrac{2}{5} & \dfrac{-\left(e^{i\theta_1}e^{-i\theta_2}\right)}{10} & \dfrac{-\left(e^{i\theta_1}e^{-i\theta_3}\right)}{10} \\[3mm] \dfrac{-\left(e^{-i\theta_1}e^{i\theta_2}\right)}{10} & \dfrac{2}{5} & \dfrac{-\left(e^{i\theta_2}e^{-i\theta_3}\right)}{10} \\[3mm] \dfrac{-\left(e^{-i\theta_1}e^{i\theta_3}\right)}{10} & \dfrac{-\left(e^{-i\theta_2}e^{i\theta_3}\right)}{10} & \dfrac{2}{5} \end{pmatrix}.$$

One can check by inspection that this matrix is positive definite. In fact, the principal minors are $\{2/5, 3/20, 1/20\}$.

The corresponding integral transform on $Gr(k, 2k + 1)$ is not injective. It has a canonical kernel. This "largest" Grassmannian over $U(2k+2)$ is endowed with an involution: $H \mapsto H^{\perp}$ which maps an affine $k + 1$ plane in affine $2k + 2$-space to its Hermitian-orthogonal complement, another $k + 1$-plane. This involution preserves the Riemannian metric and the quotient $Gr(k, 2k + 1)/\perp$ is a homogeneous Riemannian manifold covered by the Grassmannian. The involution splits functions on $Gr(k, 2k + 1)$ into odd and even parts. Each maximal totally geodesic flat is preserved by $\perp$ because any isometry Φ that moves a plane H

into $H^\perp$ belongs to a maximal torus subgroup of $U(n)$. Thus a $\perp$-odd function on $Gr(k, 2k + 1)$ has vanishing integral on each maximal flat. There is no good reason for even functions to have only vanishing integrals and in fact they do not. Invoking the representation theoretic principle above, if the maximal flat transform on $Gr(k, 2k + 1)$ has even functions in its kernel then some even function f_m must belong to the kernel. We can easily decide which f_m's are even. If a plane H in $Gr(k, 2k + 1)$ has Stiefel matrix

$$\begin{pmatrix} A \\ B \end{pmatrix},$$

where A and B are $(k + 1) \times (k + 1)$ blocks, then $H^\perp$ is spanned by the Stiefel matrix

$$\begin{pmatrix} -\overline{B} \\ \overline{A} \end{pmatrix},$$

where $\overline{(\cdot)}$ denotes complex conjugation. Inspecting the definition of f_m we see that the function f_m is $\perp$-even if and only if the integer $m_0 - m_1 + m_2 - m_3 + \cdots - \cdots + (-1)^k m_k$ is an even number. When this condition holds we can prescribe a maximal flat T with $\int_T f_m \neq 0$.

To be concrete we focus on the space $Gr(1, 3)$. We are going to integrate over the flat torus represented by the Stiefel matrix

$$m = \begin{pmatrix} e^{i\theta_1}\left(\cos(\phi) + \sin(\phi)\right) & 0 \\ \cos(\phi) & \cos(\phi) + \sin(\phi) \\ -\sin(\phi) & \cos(\phi) - \sin(\phi) \\ e^{i\theta_1}\left(\cos(\phi) - \sin(\phi)\right) & 0 \end{pmatrix}.$$

As before, it is easy to see exhibit this variety as an orbit of a maximal torus subgroup of the unitary group and hence show that this is a maximal flat submanifold. The orthonormalized version of this matrix is

$$= \begin{pmatrix} \frac{e^{i\theta_1}(\cos(\phi)+\sin(\phi))}{\sqrt{3}} & \frac{-\left(\sqrt{3}e^{i\theta_1}\cos(\phi)\right)-\sqrt{3}e^{i\theta_1}\sin(\phi)}{3\sqrt{5}} \\ \frac{\cos(\phi)}{\sqrt{3}} & \frac{2\sqrt{3}\cos(\phi)+3\sqrt{3}\sin(\phi)}{3\sqrt{5}} \\ -\frac{\sin(\phi)}{\sqrt{3}} & \frac{3\sqrt{3}\cos(\phi)-2\sqrt{3}\sin(\phi)}{3\sqrt{5}} \\ \frac{e^{i\theta_1}(\cos(\phi)-\sin(\phi))}{\sqrt{3}} & \frac{-\left(\sqrt{3}e^{i\theta_1}\cos(\phi)\right)+\sqrt{3}e^{i\theta_1}\sin(\phi)}{3\sqrt{5}} \end{pmatrix}.$$

The determinants defining f_m are the principal minors of

$$= \begin{pmatrix} \frac{2(\cos(\phi)-\sin(\phi))(\cos(\phi)+\sin(\phi))}{5} & \frac{-\left(e^{-i\theta_1}(\cos(\phi)+\sin(\phi))^2\right)}{5} \\ \frac{e^{i\theta_1}(\cos(\phi)-\sin(\phi))^2}{5} & \frac{2(\cos(\phi)-\sin(\phi))(\cos(\phi)+\sin(\phi))}{5} \end{pmatrix}.$$

The minors are, up to a positive constant, $cos^2(\phi) - \sin^2(\phi)$ and $(cos^2(\phi) - \sin^2(\phi))^2$ respectively. Since $m_1 - m_2$ is even here the integral of f_m over the flat torus is positive. It is straightforward to extend this analysis to the Grassmannian $Gr(k, 2k + 1)$.

§2. The Lagrangian Grassmannian

By restricting the full Grassmannian we can get an interesting variant of the transform in the previous section. The manifold we examine here is actually a restriction of the *real* Grassmannian. On the even dimensional Euclidean space $\mathbf{R}^{2n}$ we have the skew-symmetric quadratic form

$$\Omega = dx_1 \wedge dx_2 + dx_3 \wedge dx_4 + \ldots + dx_{2n-1} \wedge dx_{2n}.$$

We will call this the *symplectic* form. The unitary group $U(n)$ acts by linear transformations on $\mathbf{R}^{2n}$. The transformations preserve both the Euclidean inner product (because $U(n) \subset O(2n)$) and also the symplectic form (essentially the imaginary portion of the Hermitian inner product on $\mathbf{C}^n$). The symplectic form is non-degenerate on the ambient space $\mathbf{R}^{2n}$. Because of skew-symmetry, the symplectic form vanishes identically when restricted to certain n-dimensional planes such as the space of vectors whose even coordinates vanish. These planes are called *Lagrangian* planes. Because of non-degeneracy and conservation of dimension Ω does not vanish identically on any real plane of dimension larger than n. Given any two Lagrangian planes, there is a unitary linear transformation which maps one plane onto the other. The group of unitary transformations preserving the plane of "odd coordinates" above consists of orthogonal transformations in these coordinates and is isomorphic to the group O(n). The space of all Lagrangian planes in $\mathbf{R}^{2n}$ is called the *Lagrangian Grassmannian* and denoted by $\Lambda(n)$. From our remarks it is evident that this space has the presentation $U(n)/O(n)$. It may be viewed as a submanifold of the manifold of *real* n-planes in $\mathbf{R}^{2n}$.

We can also study the Lagrangian Grassmannian intrinsically. We give it the standard homogeneous Riemannian structure. Nothing could be easier than exhibiting a maximal flat torus in $\Lambda(n)$: the set of all diagonal matrices in $U(n)$ give a Stiefel-matrix presentation. But this illustrates an important difference between the complex Grassmannian and the Lagrangian Grassmannian. In the former case, the center of the unitary group (scalar unitary matrices) acts trivially, while in the latter case it acts effectively. Any flat torus in $\Lambda(n)$ is a union of orbits of the center of $U(n)$–the circle group $U(1)$. Functions on $\Lambda(n)$ can be decomposed via Fourier series into components that transform like characters under the circle group. Only functions that transform like the trivial character have a chance of being recovered from their torus integrals. This suggests that we should mod-out by the circle group and pass to the space $SU(n)/SO(n)$. This last manifold is also a symmetric space and has the added advantage of being semi-simple. In particular, its isometry group has a finite center. It is not hard to investigate the injectivity of the maximal flat transform on $SU(n)/SO(n)$. We still cannot expect to have injectivity of the maximal totally geodesic flat transform because the isometry group $SU(n)$ has a center. This center is the cyclic group of order n generated by the scalar matrix with a primitive n^{th} root of unity for eigenvalue. As before, only functions that are invariant under the action of the center have a chance of being determined by their maximal flat integrals. Looking at such functions amounts to shifting the analysis to the space $SU(n)/(SO(n) \times \mathbf{Z}_n)$ which is covered by the reduced Lagrangian Grassmannian $SU(n)/SO(n)$. The simplest way to do the analysis is to stay on the original Lagrangian Grassmannian $\Lambda(n)$ and consider only functions that are invariant under the circle group.

We need the analog of the functions f_m. Functions on $\Lambda(n)$ are spanned by the Euclidean dot products of the rows. Thus we define $f_{m_1,\dots,m_n}$ as follows:

$$
\begin{aligned}
f_{m_1,\dots,m_n} = & \ det\left[(Z^1)(Z^1)^t\right]^{m_1-m_2} \\
& \times det\left[\begin{pmatrix} Z^1 \\ Z^2 \end{pmatrix} \cdot \begin{pmatrix} Z^1 \\ Z^2 \end{pmatrix}^t\right]^{m_2-m_3} \\
& \times \cdots \times \\
& \times det\left[\begin{pmatrix} Z^1 \\ \cdot \\ \cdot \\ \cdot \\ Z^n \end{pmatrix}\begin{pmatrix} Z^1 \\ \cdot \\ \cdot \\ \cdot \\ Z^n \end{pmatrix}^t\right]^{m_n-0}.
\end{aligned}
$$

The notation here differs from the complex Grassmannian case in two ways. First, we start numbering the rows with 1 rather than 0. This is because the group of isometries consists of $n \times n$ matrices, not $(n+1) \times (n+1)$ matrices. Second, we are using matrix transform $(\cdot)^t$ rather than Hermitian adjoint $(\cdot)^*$ so as to obtain a right-$O(n)$ invariant expression. We should also note that f_m is circle-group invariant precisely when $m_1 + \dots + m_n = 0$.

The analogous working principle is valid here (see [Gr, 1985]); We will not state it explicitly. Consider now the following abelian subgroup of $U(n)$:

$$
\begin{pmatrix}
e^{\sqrt{-1}m_2\theta_2+\cdots+\sqrt{-1}m_n\theta_n} & 0 & 0 & \cdots & 0 \\
0 & e^{-\sqrt{-1}m_1\theta_2} & 0 & \cdots & 0 \\
0 & 0 & e^{-\sqrt{-1}m_1\theta_3} & \cdots & 0 \\
\vdots & \vdots & \vdots & \ddots & \vdots \\
0 & 0 & 0 & \cdots & e^{-\sqrt{-1}m_1\theta_n}
\end{pmatrix}.
$$

This is a Stiefel matrix representation of a flat totally geodesic torus of dimension $n-1$ in the Lagrangian Grassmannian. It is not a maximal flat, but after taking quotients by the center of $U(n)$ it does give a Stiefel presentation of maximal flat in the resulting manifold. Moreover, the function f_m is constant on this flat hence has non-vanishing integral. We have shown that the maximal flat transform is non-injective on $\Lambda(n)$ but becomes injective as soon as we pass to the quotient by the appropriate subgroup of the group of isometries. The situation is completely analogous to the case of the complex Grassmannians $Gr(k, 2k + 1)$ and the general theorem of [Gr, 1991].

§3. Perpendicular Transforms

In closing we present an example of a non-flat Radon transform that exhibits the same pattern of injectivity and non-injectivity that we have seen in the flat cases above. It suggests that some more general theorem about totally geodesic Radon transforms may be true.

The flat transforms above may be said to emanate from the incidence relation

$$\{(x, T) | x \text{ is a point, } T \text{ is a maximal flat, and } x \in T \}.$$

Thus the flat Radon transform integrates over the flat incidence relation. This approach leads to the double fibration formulation of Radon transforms. We will use this notion in a very informal way here: to define a Radon-like transform we merely specify the incidence relation to integrate over.

We now describe the perpendicular incidence relation on $Gr(k, n)$. Choose integers l and m with $1 \leq l \leq n$, and $m \leq min(k, n)$. We will say that a projective k-plane K is incident to a projective l-plane L in CP^n if K intersects L orthogonally in a projective m-plane (the notion of orthogonality comes from the equivalent affine picture). If $m = k < l$ then this incidence relation leads to a Radon transform that has been studied (and inverted) before ([Gr, 1986]). The general case may be of interest for its own sake and perhaps also in conjunction with I.M. Gelfand's approach to hypergeometric functions. Affine analogs of some of these transforms have been considered by Helgason [H, 1965] and Gonzalez [Go, 1987]. The first case we'll examine is $k = l = 1, m = 0, n = 3$. Thus we are looking at pairs (K, L) where K and L are projective lines in CP^3 which meet orthogonally at a projective point. In the equivalent affine picture, K and L are complex 2-dimensional subspaces of C^4 meeting orthogonally in a complex line. The Radon transform here maps functions on $Gr(1, 3)$ to functions on $Gr(1, 3)$. If $f(K)$ is such a (continuous) function then its Radon transform is the function

$$g(L) = \int_{K \in Gr(1,3)} f(K) \, d\mu(K),$$

where $d\mu$ is the appropriate rotation-invariant measure. By the working principle, we can determine the injectivity properties of this transform by simply looking at functions of the form f_m. In fact, this transform has an obvious kernel. If a 2-dimensional subspace K of C^4 meets a 2-dimensional subspace L orthogonally in a 1-dimensional subspace then it also meets $L^\perp$ orthogonally in a 1-dimensional subspace. Thus the functions $g(L)$ in the image of the Radon transform are $\perp$-even and $\perp$-odd functions have vanishing Radon transform. To see that the transform is injective on $\perp$-even functions, let K be the affine subspace of $\mathbf{C}^4$ with Stiefel matrix

$$\begin{pmatrix} 1 & 0 \\ 0 & 1 \\ 0 & 1 \\ 1 & 0 \end{pmatrix}.$$

The set of all 2-planes L incident to K is given by the following variety of Stiefel matrices:

$$\begin{pmatrix} \cos(\psi) & \cos(\phi) \\ e^{\sqrt{-1}\theta_1}\sin(\psi) & e^{\sqrt{-1}\theta_2}\sin(\phi) \\ e^{\sqrt{-1}\theta_1}\sin(\psi) & -e^{\sqrt{-1}\theta_2}\sin(\phi) \\ \cos(\psi) & -\cos(\phi) \end{pmatrix}$$

Here all angles are real. The function f_m has the value

$$[\cos^2(\psi) - \cos^2(\phi)]^{m_1-m_2} \cdot \left| e^{\sqrt{-1}\theta_2}\cos(\psi)\sin(\phi) - e^{\sqrt{-1}\theta_1}\cos(\phi)\sin(\psi) \right|^{2m_2}.$$

If f_m is a $\perp$-even function then $m_1 - m_2$ is an even integer and the integral of f_m over the incident variety is positive, hence the Radon transform is injective. All this depends on the identity $l = k = n - k$ and the involution on the Grassmannian that is available in this case. If we increase n and l by 1 we are looking at pairs (K, L) where K is a 2-dimensional subspace of $\mathbf{C}^5$, L is a 3-dimensional subspace, and K meets L orthogonally in a complex line. Here the Radon transform maps functions on $Gr(1,4)$ to functions on $Gr(2,4)$. It is no longer an endomorphism and there are no involutions in sight. Indeed, this Radon transform is injective on the entire space of continuous functions on $Gr(1,4)$. To verify injectivity, consider the point in $Gr(2,4)$ with Stiefel matrix

$$\begin{pmatrix} 1 & 0 & 0 \\ 0 & 1 & 0 \\ 0 & 0 & 1 \\ 0 & 1 & 0 \\ 1 & 0 & 0 \end{pmatrix}.$$

The Stiefel matrices incident to this one are described by the following variety:

$$\begin{pmatrix} \cos(\psi)\cos(\eta) & \cos(\phi) \\ e^{\sqrt{-1}\theta_1}\sin(\psi)\cos(\eta) & e^{\sqrt{-1}\theta_2}\sin(\phi) \\ \sin(\eta) & 0 \\ e^{\sqrt{-1}\theta_1}\sin(\psi)\cos(\eta) & -e^{\sqrt{-1}\theta_2}\sin(\phi) \\ \cos(\psi)\cos(\eta) & -\cos(\phi) \end{pmatrix}$$

Here, as before, all angles are real. The function f_m has the following value (up to a constant):

$$[\cos^2(\psi)\cos^2(\eta) - \cos^2(\phi)]^{m_1-m_2} \cdot$$
$$\left| e^{\sqrt{-1}\theta_2}\cos(\psi)\sin(\phi) - e^{\sqrt{-1}\theta_1}\cos(\phi)\sin(\psi) \right|^{2m_2}[\cos(\eta)]^{2m_2}.$$

Again, when $m_1 - m_2$ is even the integral is positive. When $m_1 - m_2$ is odd the integral is strictly increased if we remove the leftmost occurrence of the $\cos(\eta)$ factor, and the resulting integral is then zero. Hence the original integral is strictly negative.

To summarize both the flat and the perpendicular cases that we have examined, the fundamental group imposes certain restrictions on injectivity of the Radon transform and explicit calculations show that these are the only obstructions to injectivity. Hence we have a

SYMMETRY PRINCIPLE. *The Radon transform is injective whenever symmetry considerations permit.*

In [Gr, 1991] we show that this principle holds for the flat transform on any symmetric space of the compact type. We do not know at this time if the perpendicular Radon transforms fit into a general family with a corresponding injectivity result.

REFERENCES

[BGM]. M. Berger, P. Gaudichon, E. Mazet, *Le Spectre d'une variété Riemannienne*, Lecture Notes in Math no. 194, Springer Verlag, Berlin, 1971.

[D]. J. Dieudonné, *Special functions and linear representation of Lie groups*, Conf. Board Math. Sci. Series No. 42, Amer. Math. Soc., Providence, Rhode Island (1980).

[F]. P. Funk, *Über eine geometrische Anwendung der Abelschen Integralgleichung*, Math. Ann. **77** (1916), 129-135.

[Go]. Fulton Gonzalez, *Radon Transforms on Grassmann manifolds*, J. Funct. Anal. **71** (1987), 339-362.

[Gr]. E. Grinberg, *Spherical Harmonics and Integral Geometry on Projective Spaces*, Trans. Amer. Math. Soc. **279** (1983), 187-203; *On Images of Radon Transforms*, Duke J. Math. **52** (1985), 939-972; *Radon Transforms on Higher Rank Grassmannians*, J. Diff Geo. **24** (1986), 53-68; *Flat Radon Transforms on Compact Symmetric Spaces with Applications to Isospectral Deformation*, preprint (1991).

[GKK]. V.W. Guillemin, M. Kashiwara, and T. Kawai, *Seminar on Micro-Local Analysis*, Princeton University Press, 1979.

[H]. S. Helgason, *The Radon Transform on Euclidean Spaces, Two-Point Homogeneous Spaces, and Grassmann Manifolds*, Acta Math. **113** (1965); *Differential Geometry, Lie Groups, and Symmetric Spaces*, Academic Press, New York, 1978; *The X-ray transform on a symmetric space*, Proc Conf. Diff. Geometry and Global Analysis, Berlin 1979 Lecture Notes Springer Verlag **838** (1981); *Groups and Geometric Analysis*, Academic Press, New York, 1984.

[R]. J. Radon, *Über die Bestimmung von Funktionen durch ihre Integralwerte längs gewisser Manningfaltigkeiten.*, Ber. Verh. Sächs Akad. Wiss. Leipzig, Math.-Nat. kl. **69** (1917), 262-277.

[S]. R.S. Strichartz, *The explicit Fourier Decomposition of $L^2(SO(n)/SO(n-m))$*, Canad. J. Math. **27** (1975), 294-310; *Bochner Identities for Fourier Transforms*, Trans. Amer. Math. Soc. **228** (1977), 307-327; *L^p estimates for Radon transforms in Euclidean and non-Euclidean spaces*, Duke Math. J. **48** (1981), 699-727.

DEPARTMENT OF MATHEMATICS TEMPLE UNIVERSITY PHILADELPHIA, PA 19122 U.S.A.

E-mail address: grinberg@euclid.math.temple.edu

Contemporary Mathematics
Volume **140**, 1992

On Positivity Problems
for the Radon transform
and Some Related Transforms

PETER KUCHMENT

ABSTRACT. We discuss the ranges of the Radon transform, the exponential Radon, and the dual Radon transforms on the cones of non-negative functions.

§1. Introduction

Let $f(x)$ be a compactly supported function on the plane $\mathbb{R}^2$, and let

$$g(p, \omega) = (Rf)(p, \omega) = \int_{x \cdot \omega = p} f(x) \, dx$$

be its Radon transform. Here $p \in \mathbb{R}$, $\omega \in S^1 \subset \mathbb{R}^2$, and we abuse notation, denoting by dx the linear Lebesgue measure on the line $x \cdot \omega = p$.

In the X-ray tomography the function $g(p, \omega)$ represents the experimental data (see, for instance, [**13**]). The function $f(x)$ is the tissue attenuation distribution of the body, so it can not have negative values. This means that $g \in R(K)$, where K is the cone of non-negative functions. A description of the $R(K)$ could be useful, for instance, for data noise reduction.

The reader can find another interesting collection of examples of non-injectivity questions for the Radon transform (arising in mathematical economics) in the paper [**11**].

1991 *Mathematics Subject Classification.* Primary 44A12; Secondary 92C55.
Key words and phrases. Radon transform, exponential Radon transform.
Supported by the Wesley Foundation Grant #9012019, Wichita, Kansas. The Wesley Foundation is a philanthropic organization whose mission is to improve the quality of health in Kansas.
This paper is in final form and no version of it will be submitted for publication elsewhere

We encounter an analogous situation in the case of the single photon emission computed tomography (see [13]). We deal here with the exponential Radon transform

$$(R_\mu f)(p, \omega) = \int_{x \cdot \omega = p} f(x) \, exp(\mu x \cdot \omega^\perp) \, dx$$

where $\mu = const.$ is the tissue attenuation coefficient (that is assumed to be constant) and $\omega^\perp$ is the result of a 90° counterclockwise rotation of ω. The problem now is to describe $R_\mu(K)$.

The radiation dose planning problem (see [4,5]) leads to an analogous question for the dual exponential Radon transform (or " exponential back-projection"):

$$R_\mu^\# g(x) = \int_{S^1} g(x \cdot \omega, \omega) \, exp(\mu x \cdot \omega^\perp) \, d\omega$$

(here $g(p, \omega)$ is a function on the cylinder $\mathbb{R} \times S^1$). In the zero-approximation we have $\mu = 0$ here. The image $R_\mu^\# g(x) = l(x)$ is the resulting radiation dose, and $g(p, \omega)$ is the intensity of the radiation beam along the line $x \cdot p = \omega$. Because we cannot produce negative beam intensities $g(p, \omega)$, the dose $l(x)$ must belong to $R_\mu^\#(K)$ (we denote by K the cones of non-negative functions in different function spaces). We cannot, for instance, obtain the dose $l(x)$ that is the characteristic function of a region (an interior tumor). This raises the problem of description of $R_\mu^\#(K)$ (see [4,5]).

The paper is organized as follows:

Section 2 contains some simple results for the Radon transform. These results follow immediately from known facts about Radon, Fourier, and Laplace transforms.

Section 3 deals with the exponential Radon transform.

A simple relation between the direct and dual Radon transforms is described in Section 4.

Section 5 contains some partial results for the case of the dual Radon transform.

Section 6 contains final remarks.

2. Radon transform case

The Fourier slice theorem [13] connects the one-dimensional $p \to \sigma$ Fourier transform $\hat{g}(\sigma, \omega)$ of $g = Rf$ with the two-dimensional $x \to \xi$ Fourier transform $\tilde{f}(\xi)$ of $f(x)$:

(1) $$\hat{g}(\sigma, \omega) = \sqrt{2\pi} \tilde{f}(\sigma\omega).$$

We have the same relation for the Laplace transforms

$$Lg(\lambda, \omega) = \int exp(\lambda p)g(p, \omega)\, dp$$

and

$$Lf(\xi) = \int exp(\xi \cdot x)f(x)\, dx$$

(we assume that both functions f and g are bounded and compactly supported, so we have no troubles with convergence). Namely,

$$(2) \qquad\qquad (Lg)(\lambda, \omega) = (Lf)(\lambda\omega).$$

This means that knowing the data $g(p, \omega)$ we can recover by (1)-(2) the Fourier and Laplace transforms of $f(x)$. On the other hand, it is very well known how to describe the non-negativity condition in terms of Fourier or Laplace transforms. These descriptions are given by S. Bochner's and S. Bernstein's theorems, respectively (see, for instance, [2,6,15,16,18]). This enables us to obtain immediately some non-negativity results for the Radon transform.

DEFINITION 1. *A continuous function $G(\xi)$ on $\mathbb{R}^n$ is said to be positive definite if for any choice of a finite number of points $\xi_1, \dots, \xi_m \in \mathbb{R}^n$ and any vector $\eta = (\eta_1, \dots, \eta_m) \in \mathbb{C}^m$ the inequality*

$$(3) \qquad\qquad \sum_{j,k=1}^{m} G(\xi_j - \xi_k)\eta_j \overline{\eta_k} \geq 0$$

holds.

DEFINITION 2. *A distribution $T \in \mathcal{D}'(\mathbb{R}^n)$ is said to be positive definite if $T(\phi * \tilde{\bar{\phi}}) \geq 0$ for all test functions $\phi \in \mathcal{D}(\mathbb{R}^n)$, where $*$ is convolution and $\tilde{\phi}(x)$ is $\phi(-x)$.*

DEFINITION 3. *A continuous real valued function $G(\xi)$ on $\mathbb{R}^n$ is said to be exponentially convex if for any choice of a finite number of points $\xi_1, \dots, \xi_m \in \mathbb{R}^n$ and any vector $\eta = (\eta_1, \dots, \eta_m) \in \mathbb{R}^m$ the inequality*

$$(4) \qquad\qquad \sum_{j,k=1}^{m} G(\xi_j + \xi_k)\eta_j \eta_k \geq 0$$

holds.

Let us denote by K the cone of all non-negative functions $f \in C_0^\infty(\mathbb{R}^2)$.

THEOREM 1. *An even function $g(p, \omega)$ (i.e. $g(-p, -\omega) = g(p, \omega)$) belongs to $R(K)$ iff it satisfies the following conditions*
(i) $g \in C_0^\infty(\mathbb{R} \times S^1)$;

(ii) *for all* $k \in \mathbb{Z}^+$ *the function of* ω

$$\int_{-\infty}^{+\infty} p^k\, g(p,\omega)\, dp$$

is the restriction to the unit circle of a k-th order homogeneous polynomial;

(iii) *the function*

$$G(\xi) = \hat{g}(|\xi|, \xi/|\xi|), \ \xi \in \mathbb{R}^2$$

is positive definite.

This is a simple combination of Fourier slice theorem, Paley-Wiener theorem for the Radon transform [7-10], and Bochner's theorem.

THEOREM 2. *An even function* $g(p,\omega)$ *belongs to* $R(K)$ *iff it satisfies conditions i) and ii) of Theorem 1, and the condition:*

(iv) *the function*

$$G_1(\xi) = (Lg)(|\xi|, \xi/|\xi|), \ \xi \in \mathbb{R}^2$$

is exponentially convex.

Here we just apply Bernstein's theorem instead of Bochner's.

3. Exponential Radon transform case

We shall consider now the case of the exponential Radon transform R_μ. The Fourier and Laplace slice theorems (1)-(2) have simple analogs here (see [13]): if $g = R_\mu f$, then

(5) $$\hat{g}(\sigma, \omega) = \sqrt{2\pi} f(\sigma\omega + i\mu\omega^\perp)$$

(6) $$(Lg)(\lambda, \omega) = (Lf)(\lambda\omega + \mu\omega^\perp)$$

This means that knowledge of g allows us to recover the values of the Fourier transform of f on the surface

$$S_\mu = \{z = \sigma\omega + i\mu\omega^\perp | \sigma \in \mathbb{R}, \ \omega \in S^1\} \subset \mathbb{C}^2,$$

and the values of the Laplace transform Lf of f *outside* the ball $\{\xi \in \mathbb{R}^2 \,|\, |\xi| < |\mu|\}$.

The questions are now:

a) How to recognize positive definite functions, looking at their restrictions onto S_μ?

b) How to recognize positive definite functions, looking at their values outside the ball?

We do not know the answer to a) so far. The next result gives an answer to b) in our situation.

THEOREM 3. *A function $g(p, \omega)$ belongs to $R_\mu(K)$ iff it satisfies the following conditions:*

(i) $g \in C_0^\infty(\mathbb{R} \times S^1)$;

(ii) *For any odd natural number n*

$$(7) \quad \sum_{k=0}^{n} C_n^k \frac{d}{d\phi} \circ \left(\frac{d}{d\phi} - i\right) \circ \cdots \circ \left(\frac{d}{d\phi} - (k-1)i\right) \int_{-\infty}^{\infty} (\mu p)^{n-k} g(p, \omega(\phi)) \, dp$$

$$= 0,$$

where $\omega(\phi) = (\cos\phi, \sin\phi)$, and C_n^k is the binomial coefficient $\binom{n}{k}$;

(iii) *the function $G(\xi)$, defined for $|\xi| \geq |\mu|$ as*

$$(8) \qquad G(\xi) = (Lg)(\sqrt{|\xi|^2 - |\mu|^2}, (\sqrt{|\xi|^2 - |\mu|^2}\xi - \mu\xi^\perp)/|\xi|^2)$$

satisfies (4) for all $\eta \in \mathbb{R}^m$ and any choice of a finite number of points $\xi_1, \ldots, \xi_m \in \mathbb{R}^2$ such that $|\xi_j| \geq |\mu|$, and $|\xi_j + \xi_k| \geq |\mu|$ for all j, k.

(Here $\xi^\perp$ is the result of the counterclockwise $90°$ rotation of ξ.)

PROOF. By the result of [12], the conditions (i) and (ii) are equivalent to the representation $g = R_\mu f$ with some $f \in C_0^\infty(\mathbb{R}^2)$, so we have only to check that non-negativity of this function f is equivalent to the condition (iii).

Let us denote by $T_\mu(\lambda, \omega)$ the point $\lambda\omega + \mu\omega^\perp \in \mathbb{R}^2$; then T_μ is a smooth mapping from $\mathbb{R} \times S^1$ onto $\mathbb{R}^2 \setminus \{\xi \mid |\xi| < |\mu|\}$. This mapping is not one-to-one (in our two-dimensional case each point outside the ball has exactly two pre-images, in higher dimensions each such point has infinitely many pre-images). A simple geometric argument shows that

$$P_\mu(\xi) = (\sqrt{|\xi|^2 - |\mu|^2}, (\sqrt{|\xi|^2 - |\mu|^2}\xi - \mu\xi^\perp)/|\xi|^2)$$

gives a right inverse to T_μ. Because the Laplace slice theorem (6) can be written in the form $(Lg)(\lambda, \omega) = (Lf)(T_\mu(\lambda, \omega))$, we obtain that

$$(9) \qquad\qquad Lf(\xi) = Lg(P_\mu(\xi)).$$

The equality (9) shows that $G(\xi)$ in (8) is equal to $Lf(\xi)$.

We have now to show that non-negativity of $f \in C_0^\infty(\mathbb{R}^2)$ is equivalent to the condition (4) satisfied for any choice of ξ_j such that $|\xi_j| \geq |\mu|$, $|\xi_j + \xi_k| \geq |\mu|$ for all j, k. It is very easy to prove, however, that even some more restricted choice of ξ_j is sufficient. Let us choose $p_0 > |\mu|, \omega_0 \in S^1 \subset \mathbb{R}^2, 0 < \epsilon < \sqrt{2}$, and define the set

$$A = A_{p_0, \omega_0, \epsilon} = \{\xi \in \mathbb{R}^2 \mid \xi \cdot \omega_0 \geq p_0, \, |(\xi/|\xi|) - \omega_0| < \epsilon\}.$$

This is the intersection of a convex cone with a half-plane, so it is convex and homogeneous with respect to dilations with coefficients that are greater than 1. In particular, it is stable with respect to vector summation. Besides, A is disjoint with the ball of radius $|\mu|$ centered at the origin. Let us mention also that A has non-empty interior.

We are going to show now that the property (4), being satisfied only for $\xi_j \in A$, still implies non-negativity of our function $f \in C_0^\infty(\mathbb{R}^2)$.

Calculating the left hand side of (4), we get

$$(10) \qquad \int (\sum exp(\xi_j x)\eta_j)^2 f(x) \, dx$$

which is non-negative for non-negative $f(x)$. And conversely, let us assume that (10) is non-negative for all ξ_j in A. Let M be the (compact) support of f. Consider the (real) space B of linear combinations $\sum exp(\xi_j x)\eta_j$ where $\xi_j \in A$, $\eta_j \in \mathbb{R}$, $x \in M$. Because of stability of A with respect to the vector sum operation, this is a sub-algebra of $C(M)$. It has no mandatory zeros (i.e. for each point in M there exists a function from B that is not zero at that point), and separates points in M (i.e. for each pair of distinct points there exists a function with different values at those points) Both these properties easily follow from the existence of non-empty interior in A. Now the Stone-Weierstrass theorem [14] shows that B is dense in $C(M)$, and hence the set of squares of functions from B is dense in $C(M)$ in the positive cone of $C(M)$. This gives the non-negativity of f. The proof is completed.

4. A relationship between the Radon and dual Radon transforms

Let $g(p, \omega)$ be an even function from $C_0^\infty(\mathbb{R} \times S^1)$, and

$$R^\# g(x) = \int_{S^1} g(x \cdot \omega, \omega) \, d\omega, \quad x \in plane$$

be its dual Radon transform (back-projection). We shall relate $R^\# g$ to the Radon transform of some auxiliary function.

Let us denote by $Qg(x)$ the following function on the plane:

$$(11) \qquad Qg(x) = |x|^{-2} g(|x|^{-1}, x/|x|).$$

This function obviously belongs to $C^\infty(\mathbb{R}^2)$, and is $O(|x|^{-2})$ at infinity. It does not belong to $L_1(\mathbb{R}^2)$ in general (except when g is zero at $p = 0$), but it belongs to $L_p(\mathbb{R}^2)$ for any $p > 1$, and is absolutely integrable along any line.

PROPOSITION 4. *For any $g \in C_0^\infty(\mathbb{R} \times S^1)$ we have the following identity:*

$$(12) \qquad R^\# g(x) = (2\pi)^{-1}|x|^{-1}(RQg)(|x|^{-1}, x/|x|), \quad x \neq 0$$

(or, equivalently, $(RQg)(p, \omega) = 2\pi p^{-1} R^\# g(p^{-1}\omega)$).

PROOF. Let us consider the definition of $R^\# g$:

$$R^\# g(x) = \int_{S^1} g(x \cdot \omega, \omega) \, d\omega = (2\pi)^{-1} \int_0^{2\pi} g(x \cdot \omega(\phi), \omega(\phi)) \, d\phi$$

where $\omega(\phi) = (\cos \phi, \sin \phi)$. Let $y(x, \omega) = (x \cdot \omega)^{-1}\omega \in \mathbb{R}^2$. This point obviously belongs to the straight line $y \cdot x = 1$, or $y \cdot (x/|x|) = 1/|x|$. Let us denote

$w_0 := x/|x|$, $p_0 = |x|^{-1}$; then y belongs to the line $L_{p_0 w_0} = \{y \in \mathbb{R}^2 | y \cdot w_0 = p_0\}$, which is orthogonal to w_0. When ϕ runs from 0 to 2π, the point y moves along the line. It is now easy to find the relation between the angle measure $(2\pi)^{-1} d\phi$, and the linear Lebesgue measure dy on $L_{p_0 w_0}$. A simple geometric consideration shows that

$$(2\pi)^{-1} d\phi = (2\pi)^{-1} p_0 |y|^{-2} dy.$$

In other words,

$$R^{\#} g(x) = (2\pi)^{-1} p_0 \int_{L_{p_0 w_0}} |y|^{-2} g(|y|^{-1}, y/|y|) \, dy$$

$$= (2\pi)^{-1} |x|^{-1} R(Qg)(|x|^{-1}, x/|x|).$$

The proof is complete.

5. The dual Radon transform case

Let us have a function $g \in C_0^{\infty}(\mathbb{R} \times S^1)$. The question is: how to determine whether g is non-negative, looking at its back-projection $R^{\#} g$? The proposition of the previous section suggests one way of doing this: to transform our problem into a problem for the direct Radon transform. The non-negativity conditions for g and Qg are obviously equivalent because of (11) and the evenness of g, so we carry over our problem from $R^{\#}$ onto R.

THEOREM 5. *Let* $g \in C_0^{\infty}(\mathbb{R} \times S^1)$. *The following conditions are equivalent:*
(i) $g \geq 0$ *on* $\mathbb{R} \times S^1$.
(ii) *The (locally integrable) function* $F(\xi)$ *on* $\mathbb{R}^2$, *defined on* $\mathbb{R}^2 \setminus 0$ *as*

$$(13) \qquad F(\xi) = \int e^{-i|\xi|p} R^{\#} g(\xi/p|\xi|) p^{-1} \, dp$$

is positive definite in the sense of Definition 2.

The integral (13) can be understood as the one-dimensional Fourier transform with respect to the variable p from $L_q(\mathbb{R})$ into $L_{q'}(\mathbb{R}^2)$, where $q' \in [2, \infty)$. The Bochner-Schwartz theorem (see [16]), or Theorem IX.10 in [15]) says now that non-negativity of the function Qg is equivalent to the positive definiteness of the function F in the distributional sense. Let us now calculate (using (12)) the function F in terms of $R^{\#} g$. Because F is a regular distribution, it is sufficient to evaluate F for $\xi \neq 0$, which gives (up to non-essential constant factors):

$$F(\xi) = \tilde{Q}g(\xi) = (RQg)^{\wedge}(|\xi|, \xi/|\xi|) =$$

$$= (p^{-1} R^{\#} g(p^{-1}\xi/|\xi|))^{\wedge}(|\xi|) =$$

$$= \int e^{-i|\xi|p} R^{\#} g(\xi/p|\xi|) p^{-1} \, dp.$$

This gives the statement of the theorem.

Let us mention that we applied the Fourier-slice theorem, which is justified in our situation by the Lemma on page 329 of [17].

If g vanishes at $p = 0$, this gives some additional decay of $R^\# g(\xi/p|\xi|)$ at infinity, which guarantees the absolute convergence of integrals in the last theorem. In this case the function $F(\xi)$ is continuous, so we can apply the positive definiteness in the sense of Definition 1.

6. Some concluding remarks

Theorems 1 and 2 can be considered as very well known, because they represent direct combinations of known properties of Radon, Fourier, and Laplace transforms.

Concerning results off section 3, it would be interesting to understand how the property of being positive definite survives the restriction to the surface $S_\mu \subset \mathbb{C}^2$.

The relation (12) between direct and dual Radon transforms can be obtained in different ways. Our approach combines the known relation between the dual Radon transform and spherical transform (see [3]) and the standard inversion with respect to the unit circle, which is often used to unbend circles into straight lines. Another possibility is to exploit the intertwining between Euclidean and hyperbolic Radon transforms, discovered in [1]. Some considerations of the paper [17] are close to our approach.

It would be interesting to obtain some analogs of the results of Sections 4 and 5 for the case of the exponential dual Radon transform.

All results can be easily carried over to the n-dimensional case.

The practical applicability of our results has not been studied yet. There exists the possibility that they cannot be applied successfully because of numerical problems (see, for instance, the discussion of a related problem in Section 4 of [4]).

ACKNOWLEDGEMENTS

I would like to thank Professor Carlos Berenstein for helpful conversations and the referee for important comments.

References

1. C. Berenstein and E. Casadio-Tarabusi, *Range of the k-dimensional Radon transform in real hyperbolic spaces*, preprint (1991).

2. S.N. Bernstein, *Sur les fonctions absolument monotones*, Acta Math. **52** (1929), 1-66.

3. A.M. Cormack and E.T. Quinto, *A Radon transform on spheres through the origin in $\mathbb{R}^n$ and applications to the Darboux equation*, Trans. Amer. Math. Soc. **260** (1980), 575-581.

4. ———, *A problem in radiotherapy: question of non-injectivity*, Inter. J. of Imaging Sys. and Tech. **1** (1989).

5. ———, *The mathematics and physics of radiation dose planning*, Contemp. Math. **113** (1990), 41-55.

6. R. Edwards, *Functional Analysis*, Holt, Rinehart, & Winston, New York, 1965.

7. I.M. Gelfand, M.I. Graev, and N.Ya. Vilenkin, *Generalized Functions, vol. 5: Integral Geometry and Representation Theory*, Acad. Press., New York, 1964.

8. S. Helgason, *A duality in integral geometry: some generalizations of the Radon transform*, Bull. Amer. Math. Soc. **70** (1964), 435-446.

9. _____, *The Radon Transform*, Birkhäuser, Basel-Stuttgart, 1980.

10. _____S. Helgason, *Groups and Geometric Analysis*, Acad. Press., New York, 1984.

11. G.M. Henkin and A.A. Shananin, *Bernstein theorems and the Radon transform. Application to the theory of production functions*, Mathematical Problems of Tomography, Transl. Math. Monog., A.M.S., vol. 81, 1990, pp. 189-223.

12. P. Kuchment and S. Lvin, *Paley-Wiener theorem for exponential Radon transform*, Acta Appl. Math. **18** (1990), 251-260.

13. F. Natterer, *The Mathematics of Computerized Tomogaphy*, John Wiley and Sons, Chichester, 1986.

14. M. Reed and B. Simon, *Methods of Modern Mathematical Physics, vol. 1*, Acad. Press, New York, 1972.

15. _____, *Methods of Modern Mathematical Physics, vol. 2*, Acad. Press, New York, 1975.

16. L. Schwartz, *Théory des Distributions II*, Hermann, Paris, 1951.

17. D. Solmon, *Asymptotic formulas for the dual Radon transform and applications*, Math. Z. **195** (1987), 321-343.

18. P. Widder, *The Laplace Transform*, Princeton Univ. Press, Princeton, 1946.

DEPARTMENT OF MATHEMATICS AND STATISTICS WICHITA STATE UNIVERSITY WICHITA, KS 67260-0033

E-mail address: kuchment@twsuvm.bitnet

Contemporary Mathematics
Volume **140**, 1992

Cohomology Relative to
the Germ of an Exact Form

ABDELHAMID MEZIANI

0. Introduction

Let F be the germ at $0 \in \mathbf{R}^n$ of a real analytic function and let X be either $\mathcal{O}^n$, the space of germs at $0 \in \mathbf{R}^n$ of real analytic functions; $\mathcal{E}^n$ the space of germs at 0 of C^∞ functions; or $\widehat{\mathcal{O}}^n$, the ring of formal power series in n variables. The space of germs at 0 of p-forms relative to dF is defined as:

$$\Omega^p_{dF}(X) = \Omega^p(X)/(\Omega^{p-1}(X) \wedge dF) \ ,$$

where $\Omega^q(X)$ is the X-module of q-forms. The exterior derivative induces the complex

$$0 \longrightarrow X \longrightarrow \cdots \longrightarrow \Omega^{p-1}_{dF}(X) \longrightarrow \Omega^p_{dF}(X) \longrightarrow \cdots \longrightarrow \Omega^n_{dF}(X) \longrightarrow 0 \ .$$

Thus we have the cohomology groups $H^p_{dF}(X)$: the de-Rham cohomology relative to dF.

When F is the germ at $0 \in \mathbf{C}^n$ of a holomorphic function, Malgrange has proved in [3] that

$$H^p_{dF}(\mathcal{O}^n_{\mathbf{C}}) = 0 \quad \text{and} \quad H^p_{dF}(\widehat{\mathcal{O}}^n) = 0 \quad \text{for} \ p \leq q - 2,$$

where q denotes the codimension of the singular set of dF and where $\mathcal{O}^n_{\mathbf{C}}$ is the space of germs of holomorphic functions.

In this paper we are concerned with $H^1_{dF}(\mathcal{E}^n)$. We use Hironaka's theorem on resolution of singularities to show that if $F \in \mathcal{O}^n$ is real-valued, then $H^1_{dF}(\mathcal{E}^n)$ is isomorphic to a subspace of $H^1_{dF}(\widehat{\mathcal{O}}^n)$. In particular, we obtain

$$H^1(\mathcal{E}^n) = 0 \quad \text{if} \ \operatorname{codim} S(dF^c) \geq 3 \ ,$$

where F^c denotes the complexification of F and $S(dF^c)$ is the singular set of dF^c.

1991 *Mathematics Subject Classification.* Primary 35A20; Secondary 58A10.
This paper is in final form and no version of it will be submitted for publication elsewhere

The situation is more delicate when $F \in \mathcal{O}^n$ is such that $dF \wedge d\overline{F} \not\equiv 0$. Already, when dF is nonsingular, i.e. $dF(0) \neq 0$, Treves has proved in [7] that $H^1_{dF}(\mathcal{E}^n) = 0$ if and only if the level sets of F are connected. It is also proved in [4] that a generalization of this condition can be applied to the singular case for $n = 2$.

The organization of this paper is as follows. In section 1 we set the terminology and recall a version of Hironaka's theorem on resolution of singularities. In section 2 we consider the monomial case and in section 3 we prove the main result.

1. Preliminaries

In this section we set the notations, terminology, and recall a version of Hironaka's theorem on resolution of singularities.

Let V be a subset of an n dimensional real analytic manifold. We denote by $\mathcal{O}^n(V)$, $\mathcal{E}^n(V)$, and $\mathcal{F}^n(V)$ the spaces of germs along V of, respectively, real analytic, smooth (C^∞), and flat functions (a flat function along V is a C^∞ function which vanishes on V together with all its derivatives). When $V = \{0\}$ these spaces will simply be referred to as $\mathcal{O}^n$, $\mathcal{E}^n$, and $\mathcal{F}^n$. The ring of formal power series in n variables will be denoted $\widehat{\mathcal{O}}^n$. Finally, $\Omega^p(X)$ will denote the space of p-forms with coefficients in X, where X is one of the above spaces.

Let $F \in \mathcal{O}^n$. Then the exterior derivative induces the complex

$$0 \to \Omega^0(X) \to \left(\Omega^1(X)/XdF\right) \to \cdots \to \left(\Omega^n(X)/\Omega^{n-1}(X) \wedge dF\right) \to 0$$

and consequently we have the cohomology relative to dF, denoted $H^{(*)}_{dF}(X)$. An element $\eta \in \Omega^p(X)$ is said to be dF-closed if $d\eta \wedge dF = 0$ and is said to be dF-exact if there exists $\alpha \in \Omega^{p-1}(X)$ such that

$$(\eta - d\alpha) \wedge dF = 0 \ .$$

The spaces of dF-closed and dF-exact p-forms will be denoted respectively by $Z^p_{dF}(X)$ and $B^p_{dF}(X)$.

We now state a version of Hironaka's theorem on resolution of singularities (see [2] and [6]) which will be used to prove the main result.

1.1 THEOREM. *[2] Let F be a real-valued, real analytic function near $0 \in \mathbf{R}^n$, such that $F \not\equiv 0$. Then there is an open set $U \ni 0$, an n-dimensional real analytic manifold $\mathcal{M}$, and a proper analytic map $\Phi \colon \mathcal{M} \to U$ such that*

$$\Phi \colon \mathcal{M} \backslash \Phi^{-1}(F^{-1}(0)) \longrightarrow U \backslash F^{-1}(0)$$

is a diffeomorphism and such that $F \circ \Phi$ is locally a monomial. That is, for every $m \in \mathcal{M}$ there are coordinates $(x_1, \cdots, x_n)$ centred at m such that

$$F \circ \Phi = x_1^{p_1} x_2^{p_2} \cdots x_n^{p_n} \ ,$$

for some nonnegative integers $p_1, \cdots, p_n$.

2. Monomial Case

In this section we consider the case in which F is a monomial $x_1^{p_1} \cdots x_n^{p_n}$, with $p_1, \cdots, p_n$ nonnegative integers. We will assume that at least two of the integers, say p_1, p_2 are nonzero. The other cases correspond to the usual de-Rham cohomology (if $p_1 = \cdots = p_n = 0$) and the nonsingular situation (if $p_1 \neq 0$ and $p_2 = \cdots = p_n = 0$) which are known to produce the trivial cohomology. Moreover, since the cohomology relative to $x_1^{p_1} \cdots x_n^{p_n}$ is the same as that relative to $x_1^{mp_1} \cdots x_n^{mp_n}$, we can assume that $p_1, \cdots, p_n$ are relatively prime

2.1 PROPOSITION. *Let $p_1, \cdots, p_n$ be nonnegative integers. Then*

$$H^1_{d(x_1^{p_1} \cdots x_n^{p_n})}(\mathcal{F}^n) = 0 .$$

To prove this proposition, we need the following lemma.

2.1 LEMMA. *Let $f(x_i, x_j) \in \mathcal{F}^2$. Then there is $u \in \mathcal{F}^2$ such that*

(2.1) $$L_{i,j} u = f,$$

where $L_{i,j}$ denotes the vector field

$$p_j x_i \frac{\partial}{\partial x_i} - p_i x_j \frac{\partial}{\partial x_j} .$$

PROOF. Let

$$T_{x_i} f = \sum_{k \geq 0} f_k(x_j) \frac{x_i^k}{k!}$$

be the Taylor series of f with respect to x_i, i.e.

$$f_k(x_j) = \frac{\partial^k f}{\partial x_i^k}(0, x_j) \in \mathcal{F}^1 .$$

Then it is easy to verify that the function $\alpha_k \in \mathcal{F}^1$ defined by

$$\alpha_k(x_j) = \frac{-1}{p_j}|x_j|^{\frac{p_j}{p_i}k} \int_0^{x_j} \frac{f_k(\sigma)}{\sigma|\sigma|^{\frac{p_j}{p_i}k}} d\sigma$$

satisfies

$$L_{i,j}(\alpha_k(x_j)x_i^k) = f_k(x_j)x_i^k .$$

Thus from the generalized Borel theorem (see [1] for example) there exists

$$\phi \in \mathcal{F}^2 \left(\{x_j = 0\}\right)$$

such that

$$T_{x_i}\phi = \sum_{k \geq 0} \alpha_k(x_j)\frac{x_i^k}{k!} .$$

Therefore

$$f - L_{i,j}\phi \in \mathcal{F}^2 \left(\{x_i = 0\}\right) .$$

Similarly, we can find a function ψ such that

$$f - L_{i,j}\psi \in \mathcal{F}^2\left(\{x_j = 0\}\right) \ .$$

Hence,

$$\tilde{f} = f - L_{i,j}(\phi + \psi) \quad \in \mathcal{F}^2\left\{x_i x_j = 0\}\right) \ .$$

Define the function $v(x_i, x_j) \in \mathcal{F}^2(\{x_i x_j = 0\})$ by

$$v(x_i, x_j) = \frac{1}{p_i p_j} \int_{\sqrt{|cx_i^{p_i} x_j^{p_j}|}}^{|x_i|^{p_i}} \tilde{f}(\epsilon_{x_i} \sigma^{\frac{1}{p_i}}, x_j |x_i|^{\frac{p_i}{p_j}} \sigma^{-\frac{1}{p_j}}) \frac{d\sigma}{\sigma} \ ,$$

where ϵ_x denotes the sign of the nonzero real number x and c is a constant. Easily,

$$L_{i,j}v = \tilde{f} \ .$$

And thus, the function $u = v + \phi + \psi$ satisfies (2.1) $\Diamond$

PROOF OF PROPOSITION 2.1. Let

$$\eta = \sum_{j=1}^{n} \eta_j(x) dx_j \quad \in Z^1_{d(x_1^{p_1} \cdots x_n^{p_n})}(\mathcal{F}^n) \ .$$

The integrability condition

$$d\eta \wedge d(x_1^{p_1} \cdots x_n^{p_n}) = 0$$

implies that

$$(2.2) \qquad L_{i,j} f_{k,l} = L_{k,l} f_{i,j} \quad \text{for } 1 \leq i < j \leq n \text{ and } 1 \leq k < l \leq n \ ,$$

where

$$f_{i,j} = p_j x_i \eta_i - p_i x_j \eta_j$$

and the $L_{i,j}$'s are as in lemma 2.1 .

To prove that η is $d(x_1^{p_1} \cdots x_n^{p_n})$-exact, we begin with the special case

$$f_{1,2} \in \mathcal{F}^n\left(\{x_1 \cdots x_n = 0\}\right) \ .$$

In this situation the integrability condition (2.2) and lemma 2.1 imply that the function

$$u(x) = \frac{1}{p_1 p_2} \int_{\sqrt{|x_1^{p_1} \cdots x_n^{p_n}|}}^{|x_1|^{p_1}} f_{1,2}(\epsilon_{x_1} \sigma^{\frac{1}{p_1}}, x_2 |x_1|^{\frac{p_1}{p_2}} \sigma^{-\frac{1}{p_2}}, x_3, \cdots, x_n) \frac{d\sigma}{\sigma}$$

(as in lemma 2.1, ϵ_x denotes the sign of the nonzero real number x) is in the space $\mathcal{F}^n\left(\{x_1 \cdots x_n = 0\}\right)$ and satisfies

$$L_{i,j} u = f_{i,j} \quad \text{for every } i, j \leq n \ .$$

Which means

$$(\eta - du) \wedge d(x_1^{p_1} \cdots x_n^{p_n}) = 0 \ .$$

For the general case, we need only to show that an arbitrary form $\eta \in Z^1_{d(x_1^{p_1} \cdots x_n^{p_n})}(\mathcal{F}^n)$ is $d(x_1^{p_1} \cdots x_n^{p_n})$-cohomologous to a form $\tilde{\eta}$ of the special case, i.e. cohomologous to a form $\tilde{\eta}$ for which

$$\tilde{f}_{1,2} \in \mathcal{F}^n \left(\{x_1 \cdots x_n = 0\} \right) .$$

For this, let

$$T_{x'} f_{1,2} = \sum_\alpha f_{1,2}^\alpha(x_1, x_2) \frac{x'^\alpha}{\alpha!}$$

be the Taylor series of $f_{1,2}$ with respect to $x' = (x_3, \cdots, x_n)$:

$$f_{1,2}^\alpha(x_1, x_2) = \frac{\partial^{|\alpha|} f_{1,2}}{\partial x'^\alpha}(x_1, x_2, 0, \cdots, 0) \in \mathcal{F}^2 ,$$

where $\alpha = (\alpha_3, \cdots, \alpha_n)$ is an $(n-2)$-tuple of nonnegative integers,

$$x'^\alpha = x_3^{\alpha_3} \cdots x_n^{\alpha_n}, \quad \alpha! = \alpha_3! \cdots \alpha_n!, \quad \text{and } |\alpha| = \alpha_3 + \cdots + \alpha_n.$$

Let $u_\alpha \in \mathcal{F}^2$ and $\phi \in \mathcal{F}^n \left(\{x_1^2 + x_2^2 = 0\} \right)$ be such that

$$L_{1,2} u_\alpha = f_{1,2}^\alpha \quad \text{and} \quad T_{x'}\phi = \sum_\alpha u_\alpha(x_1, x_2) \frac{x'^\alpha}{\alpha!} .$$

We have then

$$T_{x'}(f_{1,2} - L_{1,2}\phi) \equiv 0 .$$

The corresponding $f'_{1,2}$ of the differential form $\eta' = \eta - d\phi$ is in the space $\mathcal{F}^n \left(\{x_3 \cdots x_n = 0\} \right)$. Let

$$T_{(x_1, x_2)} f'_{1,2} = \sum_{m,n} f_{1,2}^{\prime (m,n)}(x') \frac{x_1^m x_2^n}{m! n!}$$

be the Taylor series of $f'_{1,2}$ with respect to (x_1, x_2). Here

$$f_{1,2}^{\prime (m,n)}(x_3, \cdots, x_n) = \frac{\partial^{m+n} f'_{1,2}}{\partial x_1^m \partial x_2^n}(0, 0, x_3, \cdots, x_n)$$

belongs to $\mathcal{F}^{n-2} \left(\{x_3 \cdots x_n = 0\} \right)$.

Claim: $f_{1,2}^{\prime (m,n)} \equiv 0$ if $p_2 m = p_1 n$.

Proof of the claim. It follows from the integrability condition (2.2) that

$$L_{1,2}(T_{(x_1, x_2)} f'_{1,j}) = L_{1,j}(T_{(x_1, x_2)} f'_{1,2}) .$$

Hence, after equating coefficients in the above series, we obtain

$$m p_j f_{1,2}^{\prime (m,n)} - p_1 x_j \frac{\partial f_{1,2}^{\prime (m,n)}}{\partial x_j} = (m p_2 - n p_1) f_{1,j}^{\prime (m,n)} \quad \text{for all } m, n .$$

In particular, for $m p_2 = n p_1$, the function $f_{1,2}^{\prime (m,n)}$ is a flat solution of the linear equation

$$p_1 x_j \frac{\partial S}{\partial x_j} = m p_j S .$$

Thus, it is the trivial solution and the claim is proved.

Therefore we can write

$$T_{(x_1,x_2)}f'_{1,2} = \sum_{mp_2 \neq np_1} f'^{(m,n)}_{1,2}(x') \frac{x_1^m x_2^n}{m!n!} \ .$$

Now, if $\psi \in \mathcal{F}^n(\{x_3 \cdots x_n = 0\})$ satisfies

$$T_{(x_1,x_2)}\psi = \sum_{mp_2 \neq np_1} f'^{(m,n)}_{1,2}(x') \frac{1}{mp_2 - np_1} \frac{x_1^m x_2^n}{m!n!} \ ,$$

then

$$T_{(x_1,x_2)}(f'_{1,2} - L_{1,2}\psi) \equiv 0$$

and the form $\tilde{\eta} = \eta - d(\phi + \psi)$, which is $d(x_1^{p_1} \cdots x_n^{p_n})$-cohomologous to η, satisfies the special case. This proves the proposition $\Diamond$

REMARK 2.1. It follows from the proof of this proposition that if

$$\eta \in Z^1_{d(x_1^{p_1} \cdots x_n^{p_n})}\left(\mathcal{F}^n(\{x_{i_1} \cdots x_{i_r} = 0\})\right), \quad \text{for some } i_1, \cdots, i_r \in \{1, \cdots, n\} \ ,$$

then there is $u \in \mathcal{F}^n(\{x_{i_1} \cdots x_{i_r} = 0\})$ such that

$$(\eta - du) \wedge d(x_1^{p_1} \cdots x_n^{p_n}) = 0 \ .$$

REMARK 2.2. If v is an integral of $d(x_1^{p_1} \cdots x_n^{p_n})$ in an open neighborhood U of 0, i.e.

$$dv \wedge d(x_1^{p_1} \cdots x_n^{p_n}) = 0 \quad \text{in } U \ ,$$

then v is constant on the connected components of the level sets of $x_1^{p_1} \cdots x_n^{p_n}$. Hence, if we write

$$\mathbf{R}^n \backslash \{x_1^{p_1} \cdots x_n^{p_n} = 0\} = C_1 \cup C_2 \cup \cdots \cup C_m \ ,$$

where the C_j's are the connected components of the left hand side, then the restriction of v to C_j can be factored via $x_1^{p_1} \cdots x_n^{p_n}$. In particular, if $v \in \mathcal{F}^n$ then there are m functions $f_1, \cdots, f_m \in \mathcal{F}^1$ such that

$$v(x) = f_j(x_1^{p_1} \cdots x_n^{p_n}) \quad \text{for } x \in C_j \ .$$

For more information about factorization of integrals of differential forms see [5].

3. Main Results

In this section we prove the main results of this paper.

3.1 THEOREM. *Let $F \in \mathcal{O}^n$ be real-valued. Then $H^1_{dF}(\mathcal{E}^n)$ is isomorphic to a subspace of $H^1_{dF}(\hat{\mathcal{O}}^n)$.*

PROOF. We first prove that

$$H^1_{dF}(\mathcal{F}^n) = 0.$$

Let $\eta \in Z^1_{dF}(\mathcal{F}^n)$ and let $\Phi : \mathcal{M} \mapsto \mathbf{R}^n$ be the desingularization map of F such as in theorem 1.1 . Since $F \circ \Phi$ is monomial at each point of $\Phi^{-1}(0)$, then it follows from proposition 2.1 that $\Phi^{-1}(0)$ (which is compact) can be covered by open sets $W_1, \cdots, W_r$ such that in each one of them $\Phi^* \eta$ is $d(F \circ \Phi)$-exact. We write

$$\bigcup_{i=1}^{r} W_i \backslash (F \circ \Phi)^{-1}(0) = U_1 \cup \cdots \cup U_s ,$$

where the U_i's are the connected components of the left hand side. Moreover, since $\Phi^* \eta$ is flat on $\Phi^{-1}(0)$, then it follows from remark 2.1 that for every $i = 1, \cdots, s$ there exists

$$u_i \in C^\infty \left(\overline{U_i} \right) \bigcap \mathcal{F}^n \left(\overline{U_i} \cap \Phi^{-1}(0) \right)$$

such that

$$du_i \wedge d(F \circ \Phi) = \Phi^* \eta \wedge d(F \circ \Phi) \text{ in } \overline{U_i}.$$

Thus, for every i, j for which $\overline{U_i} \cap \overline{U_j} \neq \emptyset$,

$$u_i - u_j \in \mathcal{F}^n \left(\overline{U_i} \cap \overline{U_j} \cap \Phi^{-1}(0) \right)$$

is an integral of $d(F \circ \Phi)$. Then (remark 2.2) there are functions $f_{i,j} \in \mathcal{F}^1$ such that

$$u_i - u_j = f_{i,j} \circ F \circ \Phi \text{ in } \overline{U_i} \cap \overline{U_j}.$$

Clearly the $f_{i,j}$'s satisfy the cocycle conditions

$$f_{i,j} + f_{j,i} = 0, \quad f_{i,j} + f_{j,k} + f_{k,i} = 0.$$

Therefore, there are functions $f_i \in \mathcal{F}^1$; $i = 1, \cdots, n$, such that

$$f_i - f_j = f_{i,j} .$$

Now, the function $V \in \mathcal{F}^n(\Phi^{-1}(0))$ defined by

$$V(x) = u_i(x) - f_i(F(\Phi(x))) \quad \text{if } x \in \overline{U_i},$$

satisfies

(3.1) $$(dV - \Phi^* \eta) \wedge d(F \circ \Phi) = 0.$$

Since V is flat on $\Phi^{-1}(0)$ and satisfies (3.1), then its blowing down $v = V \circ \Phi^{-1} \in \mathcal{F}^n$ is well defined and satisfies

$$(dv - \eta) \wedge dF = 0.$$

This proves the nullity of $H^1_{dF}(\mathcal{F}^n)$.

Second, we consider the map which to an element $u \in \mathcal{E}^n$ associates its Taylor series $\widehat{u} \in \widehat{\mathcal{O}}^n$. This map induces the homomorphism

$$I : H^1_{dF}(\mathcal{E}^n) \longrightarrow H^1_{dF}(\widehat{\mathcal{O}}^n) \; ; \; I[\eta] = [\widehat{\eta}].$$

Since $H^1_{dF}(\mathcal{F}^n) = 0$, then I is injective and so $H^1_{dF}(\mathcal{E}^n)$ is isomorphic to $I(H^1_{dF}(\mathcal{E}^n))$, a subspace of $H^1_{dF}(\widehat{\mathcal{O}}^n)$. The proof of the theorem is complete $\Diamond$

The following theorem is a direct consequence of theorem 3.1 and of Malgrange's result in [3].

3.2 THEOREM. *Let $F \in \mathcal{O}^n$ be real-valued and let $\widetilde{F}$ be the complexification of F. Denote by $S(d\widetilde{F})$ the germ of the singular set of $d\widetilde{F}$. That is, the germ of the analytic variety*

$$\{z \in \mathbf{C}^n; \; d\widetilde{F} = 0\}.$$

If $\operatorname{codim} S(d\widetilde{F}) \geq 3$, *then*

$$H^1_{dF}(\mathcal{E}^n) = 0.$$

REFERENCES

1. M. Golubitsky and V. Guillemin, *Stable mappings and their singularities*, Graduate texts in mathematics, 14, Springer-Verlag, 1973.
2. H. Hironaka, *Resolution of singularities of an algebraic variety over a field of characteristic zero*, Ann. of Math. **79** (1964), 109–326.
3. B. Malgrange, *Frobenius avec singularités 1: Codimension un*, Pub. Math. I.H.E.S. **46** (1976), 163–173.
4. A. Meziani, *Condition P for singular real analytic differential forms in the plane*, The Journal of Geometric Analysis **2** (1992), 157–182.
5. _____, *On the integrals of a singular real analytic differential form in $\mathbf{R}^n$*, Trans. of the AMS **321 No 2** (1990), 583–594.
6. H. J. Sussmann, *Real-analytic desingularization and subanalytic sets: an elementary approach*, Trans. of the AMS **317 No 2** (1990), 417–461.
7. F. Treves, *On the local solvability and local integrability of systems of vector fields*, Acta Math. **151** (1983), 1–48.

DEPARTMENT OF MATHEMATICS, FLORIDA INTERNATIONAL UNIVERSITY, MIAMI, FLORIDA 33199

Contemporary Mathematics
Volume **140**, 1992

GENERALIZED CONVEX BODIES AND GENERALIZED ENVELOPES

VLADIMIR OLIKER

In [**6**] E. Lutwak introduced a useful notion of a generalized convex body. Later, the question concerning the geometric meaning of this notion was raised. The main purpose of this note is to show that a wide class of generalized convex bodies can be identified with generalized envelopes which are gradient-like maps and have been considered before. Generalized envelopes appear naturally as solutions of differential equations and in such setting they have been studied by P. Hartman and A. Wintner [**5**], Pogorelov [**11**], and other authors. Formally, they were introduced and studied in [**8**] and [**9**]. In this note, besides describing the connection between generalized convex bodies and generalized envelopes, we establish also an important property clarifying the notion of the volume of generalized convex bodies. Finally, in the last section we discuss the notion of polarity for nonconvex bodies and show its connection with generalized convex bodies and generalized envelopes.

1. Generalized convex bodies

We recall here some definitions (along with the background material) following the notation of [**6**]. Denote by $\mathcal{K}^n$ the set of convex compact bodies with nonempty interior in Euclidean space R^n. Let S^{n-1} be a unit sphere in R^n with the center at the origin O of a fixed Cartesian coordinate system. For a $K \in \mathcal{K}^n$ and any $u \in S^n$ there exists a unique supporting hyperplane to K with the outer normal u. Conversely, at every point $r \in \partial K$ there exists at least one supporting hyperplane. The "Gauss map" $\gamma : K \to S^{n-1}$ is defined as

$\gamma(r) = $ *the set of unit normals to all supporting hyperplanes at* $r \in \partial K$.

1991 *Mathematics Subject Classification*. Primary 53C45, 52A20.

The research is partially supported by NSF grant DMS-8702742 and Emory - TU-Berlin exchange program.

This paper is in final form and no version of it will be submitted for publication elsewhere.

The distance from O to the supporting hyperplane with the outer unit normal u we denote by $h_K(u)$ and refer to it as the support function of K. (The subscript K will be suppressed when there will be no danger of confusion.) For a Borel set $A \subset S^{n-1}$ the set $\gamma^{-1}(A) \subset \partial K$ is also a Borel set [2]. The surface area measure S_K of K is a Borel measure on S^{n-1} such that for any Borel set $A \subset S^{n-1}$ the measure $S_K(A) = (n-1)$-dimensional Hausdorff measure of the set $\gamma^{-1}(A)$. For K, L in $\mathcal{K}^n$ the mixed volume is given by

$$V_1(K, L) = \frac{1}{n} \int_{S^{n-1}} h_L(u) dS_K(u).$$

The Cauchy formula for the volume is $V(K) = V_1(K, K)$.

The Minkowski theorem asserts that for a given nonnegative completely additive measure S on Borel subsets of S^{n-1} such that it is not concentrated on a great sphere and

$$\int_{S^{n-1}} u dS(u) = 0$$

there exists a unique (up to a parallel translation in space) convex body whose surface area measure is S (see [B], § 8). If K and $L \in \mathcal{K}^n$ then their surface area measures satisfy the conditions in the Minkowski theorem and, therefore, for any $\alpha, \beta \geq 0$ and not vanishing simultaneously, the measure $\alpha S_K + \beta S_L$ is the surface area measure of a convex body in $\mathcal{K}^n$. This convex body (defined up to a translation in R^n) is called the Blaschke linear combination of K and L.

A generalized convex body (GCB) is a functional $\Phi : \mathcal{K}^n \to (0, \infty)$ with properties:

(i) $\Phi(K) = \Phi(TK)$, where T denotes an arbitrary parallel translation of K;

(ii) Φ is linear relative to Blaschke linear combinations;

(iii) Φ is continuous relative to the metric $|h_K - h_L|_\infty$ for $K, L \in \mathcal{K}^n$.

The collection of all GCB's will be denoted by $G\mathcal{K}^n$.

It is observed in [6] that each $K \in \mathcal{K}^n$ (more precisely, the class $\{TK\}$ where T is an arbitrary translation) can be identified with a GCB by setting

(1.1) $$\Phi_K(Q) = \frac{1}{n} \int_{S^{n-1}} h_K(u) dS_Q(u).$$

Thus in this case Φ_K is generated by the support function of the body K and, since h_K defines the body K (and is defined by K) uniquely, such identification is valid.

The **volume** $V(\Phi)$ of a $\Phi \in G\mathcal{K}^n$ is defined by putting

(1.2) $$V(\Phi)^{1/n} = inf\{\Phi(Q)/V(Q)^{(n-1)/n}\}, Q \in \mathcal{K}^n.$$

If $\Phi \in \mathcal{K}^n \cap G\mathcal{K}^n$, then (1.2) gives the usual n-volume of the body.

It follows from a result by McMullen [7] (see [6]) that for each GCB $\Phi \in G\mathcal{K}^n$ there exists a continuous function $h : S^{n-1} \to R$ such that for $K \in \mathcal{K}^n$

(1.3) $$\Phi(K) = \frac{1}{n} \int_{S^{n-1}} h(u) dS_K(u).$$

If h and h_1 are two functions corresponding to the same GCB then $h(u) = h_1(u) + (C, u)$ for some constant vector $C \in R^n$, where $(,)$ is the inner product in R^n.

The concept of a GCB is of interest in itself and it has been also useful as a technical tool in convexity theory. In [6] Lutwak used this notion to prove the upper semicontinuity of affine surface area, to define polars of nonconvex bodies, etc.

2. Generalized envelopes.

Let $h : S^{n-1} \to R$ be a continuous function. **Define a support family as a family of oriented hyperplanes $\{P_u\}$ in R^n given by the equations**

(2.1) $$P_u : (x, u) = h(u), x \in R^n, u \in S^{n-1}.$$

The collection of all support families we denote by W^n.

The function $h(u)$ in this definition gives the oriented distance from the origin O to the hyperplane P_u. In classical differential geometry **an envelope** of a support family $\{P_u\}$ is a hypersurface with the property that its family of tangent hyperplanes is $\{P_u\}$. This definition assumes implicitly that an envelope is a hypersurface with some smoothness sufficient for existence of tangent hyperplanes. However, even if h is not differentiable, it still may define a hypersurface; for example, a convex polyhedron is defined (uniquely) by its support function, that is, by its support family and neither the polyhedron nor its support function are differentiable. On the other hand h can be analytic, but the "envelope" may have singularities. For example, the function $h(u) = (C, u)$, where C is a constant vector in R^n, defines the point C as the envelope of the family $\{P_u\}$.

If in the classical definition we give up the requirement that the envelope admits tangent hyperplanes but require some smoothness from h then we can construct a wide class of hypersurfaces with interesting and useful properties [8], [9].

Let $\{P_u\}$ be a support family and assume that its generating function $h :$ $S^{n-1} \to R$ is a function of class $C^s, s \geq 1$. **A generalized envelope of class C^{s-1} is a map $r : S^{n-1} \to R^n$ given by the vector valued function**

(2.2) $$r(u) = \nabla h(u) + h(u)u, u \in S^{n-1}.$$

Here u is considered as a unit vector in R^n and ∇h is the gradient of h computed in the standard metric g of S^{n-1} induced from R^n. More explicitly, if $v^1, ..., v^{n-1}$ is a system of smooth local coordinates on $S^{n-1}, g_{jk} = (\partial u/\partial v^j, \partial u/\partial v^k), j, k = 1, ..., n-1$, is the matrix of coefficients of the metric g, and $[g^{jk}] = [g_{jk}]^{-1}$ then

$$\nabla h = \sum_{j,k=1}^{n-1} \frac{\partial h}{\partial v^j} g^{jk} \frac{\partial u}{\partial v^k}.$$

Since $(\partial u/\partial v^j, u) = 0$ for each j, the right hand side of (2.2) is a decomposition of $r(u)$ into grad $h(u)$, which is tangential to S^{n-1} at u, and $h(u)u = (r(u), u)u$ which is normal to S^{n-1} at u. Thus, each point of the hypersurface defined by $r(u)$ lies on one of the hyperplanes of the support family $(x, u) = h(u)$.

An equivalent but, perhaps, more suggestive variant of this definition (from the point of view of convexity theory) can be formulated as follows. Extend $h(u)$ so that it is defined for all $u' = \alpha u, u \in S^{n-1}, \alpha \geq 0$, by putting $H(u') = \alpha h(u)$. Then H is defined on the entire R^n and it is of class $C^s(R^n)$.

LEMMA 2.1. *Let $r(u)$ be a generalized envelope of class $C^{s-1}(S^{n-1})$, $s \geq 1$, $h(u)$ the function defining it and $H(u')$ its extension to R^n. Then*

$$grad H(u') = grad H(u')|_{u'=u} = \nabla h(u) + h(u)u = r(u).$$

Proof. We compute the $grad H(u')$ in terms of $h(u)$. In local coordinates the sphere S^{n-1} is given by $u(v) = (u_1(v^1, ..., v^{n-1}), ..., u_n(v^1, ..., v^{n-1}))$. Put $u' = (y_1, ..., y_n) = \alpha u(v), Y = [\partial v^j/\partial y_i], i = 1, ..., n, j = 1, ..., n-1$, and $U = [\partial u_i/\partial v^j]$. For convenience we identify here the metric g with its coefficient matrix, that is, $g = [g_{jk}]$.

Differentiating the relation $u' = \alpha u(v)$ with respect to y_i for $i = 1, ...n$, we get

$$(2.3) \qquad (0, ..0, 1, 0, ..., 0) = \frac{\partial \alpha}{\partial y_i} u + \alpha \sum_{j=1}^{n-1} \frac{\partial u}{\partial v^j} \frac{\partial v^j}{\partial y_i}.$$

Taking the inner product with $\partial u/\partial v^k$, we obtain $U = \alpha g Y$ and $\alpha Y = g^{-1} U$. Similarly, taking the inner product of (2.3) with u, we obtain $u_i = \partial \alpha/\partial y_i$.

Note that $grad H(u') = grad H(u')$ at $u' = u$, because $grad H(u')$ is homogeneous of order zero. Finally, differentiating H with respect to y_i at $u' = u$, we obtain

$$\frac{\partial H}{\partial y_i} = \frac{\partial \alpha}{\partial y_i} h + \alpha \sum_{j=1}^{n-1} \frac{\partial h}{\partial v^j} \frac{\partial v^j}{\partial y_i} = u_i h + \sum_{j,k=1}^{n-1} \frac{\partial h}{\partial v^j} g^{jk} \frac{\partial u_i}{\partial v^k}, i = 1, ..., n.$$

This is the assertion of the lemma in coordinate form.

Consequently, (2.2) can be written as

$$(2.4) \qquad r(u') = grad H(u'), u' \in R^n.$$

This is the map defining the generalized envelope in terms of the function $H(u')$. In convexity theory this is the traditional way of representing the boundary of a smooth convex body (see [3], p.26). For our purposes here it will be convenient

THEOREM 3.1. *Let $\Phi \in G\mathcal{K}^n$ and assume that the function h corresponding to Φ in representation (1.3) is positive. Denote by K_1 the convex body defined as the intersection of the halfspaces determined by the hyperplanes $(x, u) = h(u)$. Then $V(\Phi) = V(K_1)$.*

Proof. Denote by $h_1(u)$ the support function of K_1. Obviously, $h_1(u) \leq h(u)$. We claim that the surface area measure of K_1 vanishes on $A = \{u \in S^{n-1} | h_1(u) < h(u)\}$.

Let ω be the set of points on ∂K_1 where there exists more than one supporting hyperplane. Let $v \in A$ and $x \in \partial K_1$ be such that the supporting hyperplane through x has normal v. We want to show that $x \in \omega$. Since K_1 is the intersection of the halfspaces determined by hyperplanes $(x, u) = h(u)$, there exists a supporting hyperplane from this family through x with normal w. But $w \neq v$ because $h_1(v) < h(v)$. This means that at x there exists more than one supporting plane, and, therefore, $x \in \omega$. Then $\gamma^{-1}(A) \subset \omega$. On the other hand, it is known that $meas_{n-1}(\omega) = 0$ [2] and the first claim is proved. The above argument is adapted from [1].

We now use the formula (1.2) to compute the volume of the GCB determined by h. Note that by the preceding argument, the Cauchy formula for the volume and (1.1) $\Phi(K_1) = V(K_1)$. It follows from (1.2) that for any $K \in \mathcal{K}^n$

$$\Phi(K) \geq V(\Phi)^{1/n} V(K)^{(n-1)/n}.$$

In particular, taking $K = K_1$, we get $V(\Phi) \leq V(K_1)$. On the other hand, since $0 \leq h_1(u) \leq h(u)$, we have $V(K_1, K) \leq \Phi(K)$ for all $K \in \mathcal{K}^n$. Then

$$inf[V(K_1, K)/V(K)^{(n-1)/n}] \leq inf[\Phi(K)/V(K)^{(n-1)/n}], K \in \mathcal{K}^n.$$

The left hand side of this inequality is $V(K_1)$. Thus, $V(K_1) \leq V(\Phi)$. Consequently, $V(\Phi) = V(K_1)$. The theorem is proved.

3.2. It follows from this theorem and the example 3.1 that **neither formula (1.2) nor the Cauchy formula can be used for computing the actual volume of GCB's, if the latter is not a convex body.**

4. Polarity

Let $\mathcal{K}_0^n$ denote the subclass of $\mathcal{K}^n$ of convex bodies with the origin O in the interior. According to the classical definition of polarity (see [3], p.28) the polar body K^* of $K \in \mathcal{K}_0^n$ is given by the relation

(4.1) $K^* = \{x \in R^n | (x, y) \leq 1 \text{ for all } y \in K\}.$

Since K contains the origin O in the interior, its boundary $F = \partial K$ is starshaped relative to O and F is a graph of some continuous positive function $\rho : S^{n-1} \to (0, \infty)$. More precisely, the hypersurface F is given by the map $r(u) = \rho(u)u : S^{n-1} \to R^n$.

The function ρ is called the **radial function** of F. The function $1/\rho(u)$ is called the **distance function** of K. It is known that the body K^* is convex, its support function is $1/\rho(u)$, the distance function of K^* is the support function of K, and $(K^*)^* = K$ (see Firey, [4]).

We want to extend the notion of polarity to nonconvex compact bodies with starshaped boundaries. Denote by S_0^n the class of hypersurfaces starshaped relative to the origin with positive continuous radial functions and by W_0^n the class of support families given by $(x, u) = d(u), u \in S^{n-1}$, with $d \in C^s(S^{n-1}), s \geq 0$, and $d > 0$.

Definition. The polar of a hypersurface $F \in S_0^n$ with radial function $\rho(u), u \in S^{n-1}$, is defined as the support family F^* from W_0^n given by the equations $(x, u) = 1/\rho(u)$. The polar of a support family $P_u : (x, u) = d(u)$ from W_0^n is defined as the hypersurface $P^* \in S_0^n$ with radial function $1/d(u)$.

Clearly, this definition extends the classical one and if $F \in \mathcal{K}_0^n$ it is the same. When F is not convex but its distance function $\rho \in C^s(S^{n-1}), s \geq 1$, we can recover from its polar support family F^* the generalized envelope using formula (2.2) with $h(u) = 1/\rho(u)$. Thus, in this case we also have a geometric object representing the polar F^*. Obviously, the property $(F^*)^* = F$ is satisfied. With a slight abuse of terminology we call such generalized envelope **the polar of F** and denote it by F^*.

The analogy here with the classical case of convex bodies goes even further. Recall that the unit normal vector field on a starshaped hypersurface $F : r(u) = \rho(u)u$ is given by

$$N(u) = \frac{-\nabla \rho(u) + \rho(u)u}{\sqrt{\rho^2(u) + |\nabla(u)|^2}}$$

and its support function $z(u) = (r(u), N(u)) = \rho^2(u)/(\rho^2(u) + |\nabla\rho(u)|^2)^{1/2}$ (see, for example, [10]). The distance function of F is $1/\rho(u)$ and it is the function generating the generalized envelope F^*- the polar of F. Let r^* denote the position vector of F^* with r^* defined by (2.2). Note that $|r^*(u)| \neq 0$ for all $u \in S^{n-1}$. The hypersurface F^* may not be starshaped and then its radial function d as a function of $v = r^*/|r^*|$ is not single valued. However, it is single valued as a function of u. Then

$$d^*(u) = 1/|r^*(u)| = \rho^2(u)/(\rho^2(u) + |\nabla\rho(u)|^2)^{1/2} = z(u)$$

which is exactly the relation known for polar convex bodies [4].

It is interesting to relate the Gauss (= Gauss-Kronecker) curvatures of a hypersurface from S_0^n and of the generalized envelope polar to it (assuming, of course, that both are sufficiently smooth). Let $F \in S_0^n$ and assume that its radial function ρ is of class $C^2(S^{n-1})$. The Gauss curvature of F is given by ([10])

$$G = (\rho^2 + |\nabla\rho|^2)^{-n/2-1}\rho^{-2n+2}\frac{det[-\rho Hess\, \rho + 2\nabla\rho \otimes \nabla\rho + \rho^2 g]}{det[g]},$$

where $g, Hess$, and ∇ have the same meaning as in section 2.

The polar F^* defined by the support family $(x, u) = h(u)(= 1/\rho(u))$ has the Gauss curvature given by (see [10])

$$(4.2) \qquad G^* = \frac{det[g]}{det[Hess\ h + gh]} = \frac{1}{G}\left(\frac{\rho^2}{\rho^2 + |\nabla\rho|^2}\right)^{(n+2)/2}$$

This formula and Proposition 2.2 show that at points where G vanishes the map (2.2) defining F^* is singular. Observe that the map $r(u) = \rho(u)u$ defining F is an embedding. Note that the factor in parentheses on the right of (4.2) is equal to $(u, N(u))^2$.

For smooth convex surfaces the relation (4.2) is the basis for introducing the affine distance; see [12], p.151.

REFERENCES

1. A.D. Aleksandrov, *On the mixed volumes of convex bodies,III,*, Matem.Sb., v.3(45), no.1, 27-46, 1938
2. H. Busemann, *Convex Surfaces*, J. Wiley, New York, 1958
3. T. Bonnesen and W. Fenchel, *Theorie der Konvexen Korper*, Erg. der Mathematik und ihrer Grenzgebiete, B.3, Heft 1, Berlin, 1934
4. W. J. Firey, *Polar means of convex bodies and a dual to the Brunn-Minkowski theorem*, Canad. J. Math., 13, no.3, 444-453, 1961
5. P. Hartman and A. Wintner, *On the third fundamental form of a surface*, Amer. J. of Math., 75, 298-334, 1953
6. E. Lutwak, *Extended affine surface area*, Advances in Math., v. 85, 39-68, 1991
7. P. McMullen, *Continuous translation invariant valuations on the space of compact convex sets*, Arch. Math.34, 377-384, 1980
8. V. Oliker, *The Christoffel problem for complete surfaces*, Vestnik Leningrad University, 25, no.13, 155-158, 1970
9. V. Oliker, *Existence of complete generalized envelopes with given sum of principal radii of curvature*, Ucen. Zapiski Len. Ped. Inst., 395, 231-255, 1970
10. V. Oliker, *Hypersurfaces in R^{n+1} with prescribed Gaussian curvature and related equations of Monge-Ampere type*, Comm. in PDE's 9(8), 807-838, 1984
11. A. V. Pogorelov, *Extrinsic Geometry of Convex Surfaces*, Ch. 7, American Math. Soc., R.I., 1973
12. P. A. Schirokov and A. P. Schirokov, *Affine Differential Geometry*, Moscow 1959(In Russian)

DEPARTMENT OF MATHEMATICS AND COMPUTER SCIENCE, EMORY UNIVERSITY, ATLANTA, GEORGIA, 30030

Contemporary Mathematics
Volume **140**, 1992

A Note on Flat Radon Transforms

ERIC TODD QUINTO

ABSTRACT. Using the techniques of real analytic microlocal analysis, we prove support theorems for Radon transforms integrating with arbitrary real-analytic weights on complex and quaternion hyperplanes as well as on Cayley lines. These are flat non-overdetermined Radon transforms that satisfy the Bolker Assumption, a microlocal injectivity condition. The surfaces being integrated over are of real codimension larger than one, and a stronger support theorem is valid than Helgason's classical result. Because the sets being integrated over are planes, the geometry is simple.

1. Introduction

Radon transforms integrate or average functions over given submanifolds, and *Flat Radon transforms* are Radon transforms defined on some class of affine planes in Euclidean space. The transforms we consider here, those integrating over complex hyperplanes in $\mathbb{C}^n$, quaternion hyperplanes in $\mathbb{H}^n$ and on Cayley lines, are all flat. The double fibration [**GGS**] allows one to define Radon transforms over fairly general sets and with general weights, and Guillemin and Sternberg (*e.g.,* [**GS**]) have used this framework and microlocal analysis to understand properties of Radon transforms (see also [**GU 1989**]). These advances make it natural to investigate Radon transforms with arbitrary weights.

A support theorem for a Radon transform provides restrictions on the support of a function from information on the support of its Radon transform. Classical techniques have been used to prove support theorems for the transforms with canonical measures on hyperplanes in $\mathbb{R}^n$ [**He**] and $\mathbb{C}^n$ [**GGV**]. The problem is more subtle for transforms with general weights, and support theorems have

1991 *Mathematics Subject Classification.* Primary: 44A12, Secondary: 58G15.

Key words and phrases. Radon transform, support theorem, microlocal analysis, Fourier integral operator.

This work has benefited from discussions with Fulton Gonzalez and Eric Grinberg and from the gracious hospitality of Alfred Louis and Universität des Saarlandes. The author was partially supported by the Humboldt Stiftung and NSF grant MCS 8901203.

This paper is in final form and no version of it will be submitted for publication elsewhere.

been proven under either strong symmetry conditions such as rotation invariance (*e.g.*, [Q 1987], [Or]) or strong smoothness conditions on the weight such as real-analyticity (*e.g.*, [BQ 1987]).

In this article, we will give a general support theorem, Theorem 1.1, that is valid for flat Radon transforms with real analytic weights. One can use the theorem to get inside the convex hull of the support of the function being integrated over; this is possible because the Radon transforms we consider integrate over planes of real codimension greater than one. The key to our proofs is the fact that the transforms we consider all satisfy the *Bolker Assumption*, a microlocal injectivity condition on the double fibration that defines them (Definition 2.1). If the Bolker Assumption does not hold for a Radon transform then more subtle geometric arguments (*e.g.*, [BQ 1992]) are needed to prove strong support theorems. We consider the geometrically simple cases of (1.1) in order to demonstrate the fundamental ideas behind our techniques. Our goal is to demonstrate how easy it is to prove support theorems using microlocal techniques when the Bolker Assumption holds.

In general, X denotes the ambient space functions are defined on and Y denotes the set of submanifolds of X that the Radon transform integrates over. In our case, X will be an Euclidean space and Y will be a set of planes of fixed codimension. In order to simplify the discussion, we will consider only the classical non-overdetermined cases:

(1.1)
$$\begin{aligned} \text{for } X = \mathbb{C}^n : \quad &Y \text{ is the set of complex hyperplanes} \\ \text{for } X = \mathbb{H}^n : \quad &Y \text{ is the set of quaternion hyperplanes} \\ \text{and for } X = \mathcal{C}^2 : \quad &Y \text{ is the set of Cayley lines} \end{aligned}$$

where $\mathcal{C}^2$ is the Cayley plane. In each of these cases, the submanifolds in Y have codimension larger than one. We do not consider the transform on real hyperplanes in $\mathbb{R}^n$ because Theorem 1.1 was proven for that case in [BQ 1987].

All non-overdetermined Radon transforms that satisfy the Bolker Assumption are in certain ways quite similar to the classical flat ones (or the rank-one horocycle transform). In particular, each manifold in Y must have real codimension in X of either 1, 2, 4, or 8, and R_μ^* must integrate over sets similar to the ones for these examples [Q 1981]. A stronger correspondence holds in the complex category [Ja].

The *incidence relation* of R_μ is the set $Z = \{(x, H) \in X \times Y \,|\, x \in H\}$ [He]. Let the weight $\mu(x, H)$ be real analytic and nowhere zero on Z. Let $f \in C_c(X)$, then the Radon transform of f is defined for $H \in Y$ by

(1.2)
$$R_\mu f(H) = \int_{x \in H} f(x)\mu(x, H)dm_H(x)$$

where dm_H is the canonical measure on H. Our main theorem is:

THEOREM 1.1. *Let X and Y be given in* (1.1). *Assume the Radon transform $R_\mu : \mathcal{E}'(X) \to \mathcal{E}'(Y)$ has nowhere zero real analytic weight μ. Let $\mathcal{A}$ be an open* *connected subset of Y. Assume $f \in \mathcal{E}'(X)$ with $R_\mu f(H) = 0$ for all $H \in \mathcal{A}$ and assume, for some $H_o \in \mathcal{A}$, H_o is disjoint from supp f. Then for all $H \in \mathcal{A}$, H is disjoint from supp f.*

An example in [**HSSW**] demonstrates the necessity of the assumption that some plane in $\mathcal{A}$ is disjoint from supp f. Using local versions of our arguments, one can easily prove local support theorems similar to those in [**Gl**], [**Q 1991**, Theorem 4.3]. This support theorem for classical measures on complex lines in $\mathbb{C}^2$ is in [**Gl 1991**].

In §2, the microlocal analysis of R_μ will be given, and the main theorem will be proven in §3. Remark 3.2 provides an outline of the proof of Theorem 1.1 more generally.

2. Flat Radon transforms and the Bolker Assumption

In this section we specify what singularities are detected by R_μ. The relevant diagrams are the double fibration (below left) [**GGS**] and the corresponding diagram on the cotangent spaces (below right):

$$
\begin{array}{ccc}
Z \xrightarrow{\pi_Y} Y & \qquad & \Gamma = N^*(Z) \setminus 0 \xrightarrow{\pi_Y} T^*(Y) \\
\end{array}
$$

(2.1)
$$
\begin{array}{ccc}
\Big\downarrow{\pi_X} & \qquad & \Big\downarrow{\pi_X} \\
X & & T^*(X)
\end{array}
$$

In (2.1), N^*Z is the conormal bundle of Z in $T^*(X \times Y)$. And, in general, if A is a submanifold of B then we will let N^*A denote the conormal bundle of A in T^*B.

DEFINITION 2.1. *If the Radon transform R_μ is defined by the double fibration, (2.1), then R_μ satisfies the **Bolker Assumption** if and only if $\pi_Y : \Gamma \to T^*Y$ is an injective immersion.*

PROPOSITION 2.2. *The Radon transforms for the manifolds in (1.1) all satisfy the Bolker Assumption. Let R_μ be a Radon transform with nowhere zero real analytic weight μ that satisfies the Bolker Assumption. Let $f \in \mathcal{E}'(X)$ and let $H_1 \in Y$. Assume $R_\mu f(H) = 0$ for all H in an open neighborhood of H_1. Let $(x, \eta) \in N^*H_1 \setminus 0$. Then $(x, \eta) \notin \mathrm{WF}_A(f)$.*

PROOF OF PROPOSITION 2.2. We will show below that these Radon transforms all satisfy the Bolker Assumption. It is well known [**GS**] that if a Radon transform satisfies the Bolker Assumption, then R_μ is an elliptic Fourier integral operator associated with the Lagrangian manifold Γ and $R_\mu^* R_\mu$ is an elliptic pseudodifferential operator. The statements of Proposition 2.2 follow immediately from the calculus of analytic elliptic Fourier integral operators [**SKK**], [**Ka**]: As $R_\mu f(H) = 0$ for H near H_1, then $(\mathrm{WF}_A R_\mu f) \cap (T_{H_1}^* Y \setminus 0) = \emptyset$. *Using the calculus of elliptic Fourier integral operators one sees that* $\left[\pi_X \circ \pi_Y^{-1}(T_{H_1}^* Y \setminus 0) \right] \cap \mathrm{WF}_A f = \emptyset$. *But* $\pi_X \circ \pi_Y^{-1}(T_{H_1}^* Y \setminus 0) = N^*H_1 \setminus 0$ *(see e.g., [**Q 1991**, Proposition 3.2] or check using the local coordinates given below).*

For each case in (1.1) it is straightforward to check that the Bolker Assumption holds, and this is known, at least in the folklore. The complex case was done in [**Q 1978**]. We will sketch the proof for the other two cases.

Unless otherwise indicated, boldface letters will denote quaternion objects and italic letters will denote real objects. If $\mathbf{1}$, $\mathbf{i}$, $\mathbf{j}$, and $\mathbf{k}$ are the standard basis for $\mathbb{H}$ then for $\mathbf{p} \in \mathbb{H}$ we let $\mathbf{p} = p^r\mathbf{1}+p^i\mathbf{i}+p^j\mathbf{j}+p^k\mathbf{k}$ where p^r, p^i, p^j, and p^k are real and $\overline{\mathbf{p}} = p^r\mathbf{1}-p^i\mathbf{i}-p^j\mathbf{j}-p^k\mathbf{k}$. If $\mathbf{x} \in \mathbb{H}^n$, then we write $\mathbf{x} = x^r\mathbf{1}+x^i\mathbf{i}+x^j\mathbf{j}+x^k\mathbf{k}$ where x^r, x^i, x^j, and x^k are in $\mathbb{R}^n$. The real coordinates of $\mathbf{x}$ are denoted similarly; for example, x_ℓ^r is the ℓth real coordinate of $x^r \in \mathbb{R}^n$. This gives an identification of $\mathbb{H}^n$ with $\mathbb{R}^{4n}$, $\mathbf{x} \to (x^r, x^i, x^j, x^k)$. The standard real basis vectors are denoted: e_ℓ^r, etc. Note that e_n^r is identified with the basis vector $\mathbf{e}_n \in \mathbb{H}^n$. Let $x \cdot y$ be the real inner product of x and y under this identification.

The quaternion inner product on $\mathbb{H}^n$ is denoted $\langle \mathbf{x}, \mathbf{y} \rangle = \sum_{\ell=1}^n \mathbf{x}_\ell \overline{\mathbf{y}}_\ell$. The quaternion hyperplane is defined by $H(\boldsymbol{\omega}, \mathbf{p}) = \{\mathbf{x} \in \mathbb{H}^n | \langle \mathbf{x}, \boldsymbol{\omega} \rangle = \mathbf{p}\}$ for $(\boldsymbol{\omega}, \mathbf{p}) \in S^{4n-1} \times \mathbb{H}$. Also, $H(\boldsymbol{\omega}, \mathbf{p}) = H(\mathbf{c}\boldsymbol{\omega}, \mathbf{p}\overline{\mathbf{c}})$ for all $\mathbf{c} \in \mathbb{H}$ with $\mathbf{c}\overline{\mathbf{c}} = 1$.

To show π_Y is an injective immersion it is sufficient to choose $H_o \in Y$ and show

$$(2.2) \qquad \pi_Y : \pi_Y^{-1}[T_{H_o}^*Y \setminus 0] \to T_{H_o}^*Y \setminus 0$$

is an injective immersion. This is true as the projection $Z \to Y$ is a fibration and so π_Y can fail to be an immersion only "above the fibers of $Z \to Y$," not transversal to them. By a rotation in $\mathbb{H}^n$, we can assume $H_o = H(\mathbf{e}_n, p_o)$ for some $p_o \in \mathbb{R}$. Now choose real local coordinates near H_o : $\tilde{Y} = \{(\boldsymbol{\omega}, p) \in S^{4n-1} \times \mathbb{H} | \omega_n^r > 0, \omega_n^i = 0 = \omega_n^j = \omega_n^k\}$, and use the corresponding coordinates for $\tilde{Z} \subset Z$.

For $x \in \mathbb{R}^{4n}$, let $\mathcal{I}x$ be the image of $\mathbf{i}\mathbf{x}$ under the above identification of $\mathbb{H}^n$ with $\mathbb{R}^{4n}$, $x \to \mathbf{x} \to \mathbf{i}\mathbf{x} \to \mathcal{I}x$. Define $\mathcal{J}x$ and $\mathcal{K}x$ analogously. Then $\tilde{Z} \subset X \times \tilde{Y}$ is given by the real valued equations

$$(2.3) \qquad x \cdot \omega = p^r, \ x \cdot \mathcal{I}\omega = p^i, \ x \cdot \mathcal{J}\omega = p^j, \ x \cdot \mathcal{K}\omega = p^k.$$

The total differentials of these functions form a basis for fibers of $N^*\tilde{Z}$. Thus if $\Gamma_o = \pi_Y^{-1}(T_{(e_n^r, p_o)}^*Y)$ is the set of fibers of Γ above $\{(x, e_n^r, p_o) | \langle \mathbf{x}, \mathbf{e}_n \rangle = p_o\}$, then Γ_o is given by

$$(2.4) \qquad \begin{aligned} &\big(x, e_n^r, p_o; \ \alpha_1(e_n^r\mathbf{dx} + (Px)\mathbf{d\omega} - \mathbf{dp}^r) + \alpha_2(\mathcal{I}e_n^r\mathbf{dx} - P(\mathcal{I}x)\mathbf{d\omega} - \mathbf{dp}^i) \\ &\quad + \alpha_3(\mathcal{J}e_n^r\mathbf{dx} - P(\mathcal{J}x)\mathbf{d\omega} - \mathbf{dp}^j) + \alpha_4(\mathcal{K}e_n^r\mathbf{dx} - P(\mathcal{K}x)\mathbf{d\omega} - \mathbf{dp}^k)\big) \end{aligned}$$

where the real constants α_1, α_2, α_3, and α_4 are not all zero. Here $e_n^r\mathbf{dx}$ indicates the cotangent vector in $T_x^*\mathbb{R}^{4n}$ corresponding to e_n^r. Also, P is the orthogonal projection onto the plane $\{x \in \mathbb{R}^{4n} | x_n^r = x_n^i = x_n^j = x_n^k = 0\}$, so $Px\mathbf{d\omega}$ is the appropriate covector in $T_{e_n^r}^*\tilde{S}$. In (2.4) we have used that the operators $\mathcal{I}$, $\mathcal{J}$, and $\mathcal{K}$ are skew symmetric. Now $\pi_Y : \Gamma_o \to T_{(e_n^r, p_o)}^*\tilde{Y}$ is the projection onto the $\mathbf{d\omega}$, $\mathbf{dp}^r$, $\mathbf{dp}^i$, $\mathbf{dp}^j$ and $\mathbf{dp}^k$ cotangent coordinates of (2.4). Thus π_Y is an injective immersion because x is given smoothly by

$x = p_o e_n^r + \frac{(\alpha_1 I + \alpha_2 \mathcal{I} + \alpha_3 \mathcal{J} + \alpha_4 \mathcal{K}) P(\alpha_1 x - \alpha_2 \mathcal{I} x - \alpha_3 \mathcal{J} x - \alpha_4 \mathcal{K} x)}{\alpha_1^2 + \alpha_2^2 + \alpha_3^2 + \alpha_4^2}$. This is true because P
commutes with $\mathcal{I}$, $\mathcal{J}$, and $\mathcal{K}$ (the operators can be viewed as simple operators
on $\mathbb{H}^n$). This finishes the quaternion calculation.

The Cayley numbers, $\mathcal{C}$, are defined as an alternative division algebra over $\mathbb{R}$
in [**Sc**, p. 48, p. 45]. Each number $c \in \mathcal{C}$ will be written $c = c_1 + v c_2$ where
c_1 and c_2 are in $\mathbb{H}$ and v is the square root of -1 that is added to $\mathbb{H}$ to form
$\mathcal{C}$. Because Cayley numbers are not associative, the local parameterization of
Cayley lines:

$$(2.5) \qquad \mathcal{C}^2 \ni (c, d) \longrightarrow \ell : y = cx + d$$

will be used (see [**Hi**]). Thus, the two quaternion equations:

$$(2.6a) \qquad \mathbf{y}_1 - \mathbf{c}_1 \mathbf{x}_1 + \mathbf{x}_2 \overline{\mathbf{c}}_2 - \mathbf{d}_1 = 0$$
$$(2.6b) \qquad \mathbf{y}_2 - \overline{\mathbf{c}}_1 \mathbf{x}_2 - \mathbf{x}_1 \mathbf{c}_2 - \mathbf{d}_2 = 0$$

locally define Z. Now write these equations as eight real equations (in order:
real part of (2.6a), imaginary part of (2.6a),...., "**k**" part of (2.6b)) and take their
differentials. This gives an expression for Γ_o analogous to (2.4) but with eight
real constants, $\alpha_1, \ldots, \alpha_8$ satisfying $\alpha_1^2 + \cdots + \alpha_8^2 \neq 0$. Now take the projection,
π_Y, onto $T^*_{(c,d)} Y$. The constants $\alpha_1, \ldots \alpha_8$ are given by the $\mathbf{dd}_1^r$, $\mathbf{dd}_1^i, \ldots, \mathbf{dd}_2^k$
coordinates, respectively, of this projection.

Let $\mathbf{A} = \alpha_1 1 + \alpha_2 \mathbf{i} + \alpha_3 \mathbf{j} + \alpha_4 \mathbf{k}$ and let $\mathbf{B} = \alpha_5 1 + \alpha_6 \mathbf{i} + \alpha_7 \mathbf{j} + \alpha_8 \mathbf{k}$. Showing
π_Y is an injective immersion is now easily seen to be equivalent to solving the
two linear quaternion equations:

$$(2.7) \qquad \begin{aligned} -\mathbf{A}\overline{\mathbf{x}}_1 - \mathbf{x}_2 \overline{\mathbf{B}} &= \mathbf{E}_1 \\ -\overline{\mathbf{x}}_1 \mathbf{B} + \overline{\mathbf{A}} \mathbf{x}_2 &= \mathbf{E}_2 \end{aligned}$$

for $\mathbf{x}_1$ and $\mathbf{x}_2$ where $\mathbf{E}_1$ is the quaternion equivalent to the $\mathbf{dc}_1$ coordinate of
this projection, and $\mathbf{E}_2$ comes from the $\mathbf{dc}_2$ coordinate. Equations (2.7) are easy
to solve using the fact that $\overline{\mathbf{A}}\mathbf{A} + \mathbf{B}\overline{\mathbf{B}} = \alpha_1^2 + \cdots + \alpha_8^2$ is a positive real number.
This finishes the proof of Theorem 2.1. $\blacksquare$

3. Proof of Theorem 1.1

PROOF. The arguments are related to those in [**BQ 1987**] for the hyperplane
transform. We will use planes in $\mathcal{A}$, Proposition 2.2, and Lemma 3.1 below to eat
away at supp f. To use Lemma 3.1 one needs codimension one surfaces. Since
the Radon transforms from (1.1) all integrate over planes of codimension greater
than one, we will construct codimension one surfaces out of planes $H \subset \mathcal{A}$. Then
we will use these surfaces to eat away at supp f.

Let $H_o \in \mathcal{A}$ be as in the statement of Theorem 1.1 and assume the conclusion
of the theorem is false. Let $H_2 \in \mathcal{A}$ be such that $H_2 \cap$ supp $f \neq \emptyset$, and let
$s : [0, 1] \to \mathcal{A}$ be a continuous path from H_o to H_2. Now choose $\epsilon > 0$ so that:

(3.1) if $x \in X$ and x is within ϵ units of H_o, then $x \notin$ supp f.

(3.2) if $H_1 \in Y$ is both parallel some plane $H \in s([0,1])$ and within ϵ units of H, then $H_1 \in \mathcal{A}$;

This is possible because H_o is disjoint from supp f and $\mathcal{A}$ is open. If $t \in [0,1]$ define $T(t)$ to be the cylinder generated by all $H \in Y$ of parallel to the plane $s(t)$ and ϵ units from $s(t)$. This construction can be done because Y contains all planes parallel any given plane in Y. By (3.1), $T(0)$ is disjoint from supp f and by assumption, $T(1)$ meets supp f. Let t_1 be the smallest value of $t \in [0,1]$ such that $T(t)$ meets supp f. By the choice of t_1, $T_1 = T(t_1)$ meets supp f only on the boundary, ∂T_1. Let $x \in \partial T_1 \cap$ supp f. Then by (3.2) the plane $H_1 \subset T(t_1)$ that contains x is in $\mathcal{A}$.

Let $\eta \in N_x^* T_1 \setminus 0$; then, by the construction of T_1, $\eta \in N_x^* H_1$. This and Proposition 2.2 imply that $(x, \eta) \notin \mathrm{WF}_A(f)$. A theorem of Hörmander, Kawai, and Kashiwara is the final key to the proof:

LEMMA 3.1. [**Hö**, Theorem 8.5.6], [**Ka**] *Let* $f \in \mathcal{E}'(X)$ *and assume* f *is zero on the interior of* T_1. *If* $x \in \partial T_1 \cap$ supp f *and* $(x, \eta) \in N^*(\partial T_1) \setminus 0$, *then* $(x, \eta) \in \mathrm{WF}_A(f)$.

This contradicts the assumption that $T(1)$ meets supp f. ∎

Remark 3.2. Theorem 1.1 is valid for many other Radon transforms, and the author is now generalizing this proof to arbitrary Radon transforms satisfying the Bolker Assumption. In general, the simple construction of $T(t)$ above is impossible, but under some geometric assumptions, the above proof can be used. The proof will now be sketched for such a case: the full $k-$plane transform with nowhere zero real analytic weight, R_μ. Using arguments related to those in Proposition 2.2 one can prove that the Bolker Assumption holds for R_μ. One chooses $s(t)$, ϵ, and $T(t)$ as in the proof of Theorem 1.1, and one finishes the proof as above. This equivalent construction of $T(t)$ is possible because the $k-$planes parallel any given $k-$plane are all in Y.

REFERENCES

BQ 1987. Boman, J. and Quinto, E., *Support theorems for real analytic Radon transforms*, Duke Math. J. **55** (1987), 943–948.

BQ 1992. _____, *Support theorems for real analytic Radon transforms on line complexes in* $\mathbb{R}^3$, Trans. Amer. Math. Soc. (to appear).

GGS. Gelfand, I., Graev, M., and Shapiro, Z., *Differential forms and integral geometry*, Functional Anal. and Appl. **3**, 101–114.

GGV. Gelfand, I., Graev, M. and Vilinkin, N., *Generalized Functions, Vol. 5*, Academic Press New York, New York.

Gl 1991. Globevnik, J., *A boundary Morera Theorem in* $\mathbb{C}^2$, preprint (1991).

Gl 1992. _____, *A support theorem for the X-ray transform*, J. Math. Anal. Appl. **165** (1992), 284–287.

GU 1989. Greenleaf, A., and Uhlmann, G., *Non-local inversion formulas in integral geometry*, Duke J. Math. **58** (1989), 205–240.

GS. Guillemin, V. and Sternberg, S., *Geometric Asymptotics*, Amer. Math. Soc., Providence, RI, 1977.

HSSW. Hamaker, C., Smith, K., Solmon, D., and Wagner, S., *The Divergent Beam X-ray Transform*, Rocky Mt. J. Math. **10** (1980.), 253–283.

He. Helgason, S., *The Radon transform on Euclidean spaces, compact two-point homogeneous spaces, and Grassmann manifolds*, Acta Math. **113** (1965), 153–180.

Hi. Hilton, P. J., *An Introduction to Homotopy Theory*, Cambridge University Press, Cambridge, England, 1961.

Hö. Hörmander, L., *The Analysis of Linear Partial Differential Operators I*, Springer, New York, 1983.

Ja. Jamshidian, F., *Integral geometry on plane complexes*, Ph.D. Dissertation, Harvard University, 1980.

Ka. Kaneko, A., *Introduction to Hyperfunctions*, Kluwer, New York, 1989.

Or. Orloff, J., *Invariant Radon transforms on a symmetric space*, Contemporary Math. **113** (1990), 233–242.

Q 1978. Quinto, E., *On the locality and invertibility of Radon transforms*, Ph.D. Thesis, MIT.

Q 1981. _____ Topological restrictions on double fibrations and Radon transforms, Proc. Amer. Math. Soc. **81** (1981), 570–574.

Q 1987. _____, *The injectivity of rotation invariant Radon transforms on complex hyperplanes in $\mathbb{C}^n$*, Contemporary Math. **63**, 245–260.

Q 1991. _____, *Real-analytic Radon transforms on rank one symmetric spaces*, Proc. Amer. Math. Soc. (to appear).

SKK. Sato, M., Kawai, T., and Kashiwara, M., *Hyperfunctions and pseudodifferential equations*, Lecture Notes in Math., vol. 287, Springer Verlag, New York, 1973, pp. 265–529.

Sc. Schafer, D., *Introduction to Nonassociative Algebras*, Academic Press, New York, 1966.

DEPARTMENT OF MATHEMATICS, TUFTS UNIVERSITY, MEDFORD, MA 02155 USA

E-mail address: equinto@math.tufts.edu

Contemporary Mathematics
Volume **140**, 1992

Self–similarity on Nilpotent Lie Groups

ROBERT S. STRICHARTZ

ABSTRACT. Various aspects of geometric analysis on Euclidean space, involving the notion of self–similarity, are generalized to classes of nilpotent Lie groups with dilations. In particular, we discuss the existence of self–similar tilings, the properties of Hausdorff measures and other fractal measures, spectral asymptotics of self–similar measures and distributions, and multi–periodic functions.

§1 Introduction

Self–similarity is an important concept in geometric analysis on Euclidean space ([F2], [H], [M]). The basic structures involved are the translations and dilations, with orthogonal transformations and lattice subgroups also playing a subsidiary role. There are several classes of nilpotent Lie groups which possess analogous structures. The goal of this paper is to extend some of the Euclidean space self–similar analysis to this wider context.

We begin with the notion of self–similar tiling. Consider the standard tiling of $\mathbb{R}^n$ by unit cubes. Each tile is a translate of a single tile by an element of the lattice subgroup $\mathbb{Z}^n$, and if we dilate a tile by a factor of 2, each expanded tile contains exactly 2^n tiles. Other, more exotic self–similar tilings, in which the boundaries of the tiles are fractal, have been constructed ([B], [GH], [K], [Th]). We will consider a class of nilpotent Lie groups which have both dilations and lattice subgroups, and show how to construct the analogue of the cubic tiling which is self–similar. It turns out that in order to do this we are forced to create tiles with fractal boundaries. For simplicity of exposition we first give the construction, in §2, for a class of Heisenberg type groups. In §3 we extend the construction to what seems to be the most general class of groups (rational graded nilpotent Lie groups) for which it makes sense. We have not attempted

1991 *Mathematics Subject Classification.* 43A80, 28A80.
Research supported in part by the National Science Foundation, grant DMS–9103348.
This paper is in final form and no version of it will be submitted for publication elsewhere.

a general description of all self–similar tilings for these groups, but that would seem to be a reasonable next goal.

Because the cubic tiling of Euclidean space is such a widely used tool, there are numerous applications of our analogous construction. Often, the proof involves nothing more than observing that the Euclidean proof extends with minor modifications. We discuss some of these applications in §4, concentrating on maximal functions and properties of Hausdorff measures. It must be emphasized that the use of our self–similar tilings may not be essential to obtaining these results, since analysis on nilpotent Lie groups has advanced quite far without them.

The remainder of the paper is devoted to extending a theory that might be called "fractal spectral asymptotics" to nilpotent Lie groups. The theory, developed in [S1], [S2], [S3], [S4], [L], [LW], and [JRS], relates the asymptotic behavior of spectral expansions of fractal objects to the fractal properties of the objects. In §5 we consider general fractal measures μ on the group and the asymptotics of the heat semigroup expansion $e^{t\mathcal{L}}\mu$, where $\mathcal{L}$ is a sub–Laplacian on the group. In §6 we specialize to self–similar measures. These are generalizations of the familiar Cantor measure on the line. We show how to extend the construction of self–similar distributions in [S3] to nilpotent Lie groups of step two, using the Euclidean rather than the group Fourier transform. This construction is important because in certain special cases it is used in the theory of wavelets [DL]. We hope that the ideas developed in this paper can be used to develop a wavelet theory analogous to [S5]. A different approach to wavelet theory on nilpotent Lie groups is given in [Le].

In Euclidean space, the Fourier transforms of self–similar measures and wavelets exhibit a structure involving both additive and multiplicative periodicity. For example, the Fourier transform of the Cantor measure is $F(x) = \Pi_{k=1}^{\infty} \cos \pi x / 3^k$. In §7 we describe a class of multi–periodic functions on nilpotent Lie groups with similar structure, and extend some of the results in [JRS] to this context. Of course, in this context we do not have a Fourier analytic interpretation for these functions.

For the basic facts and definitions concerning nilpotent Lie groups we refer the reader to Folland and Stein [FS]. All the Lie groups we consider will be graded, meaning they possess a dilation structure. Various other hypotheses are added when needed.

Since many of the proofs we give are minor modifications of proofs given elsewhere, we will occasionally omit proofs or give abbreviated summaries.

§2. Tilings on Heisenberg Groups

In this section we give the basic tiling construction for a special class of groups of "Heisenberg" type. The goal is to set forth the basic simple ideas involved, without the distraction of the general setting. We assume $G = \mathbb{R}^{n_1} \times \mathbb{R}^{n_2}$ with the group law

$$(2.1) \qquad (x, y) \circ (x', y') = (x + x', y + y' + S(x, x'))$$

where $x \in \mathbb{R}^{n_1}$, $y \in \mathbb{R}^{n_2}$ and $S(x, x')$ is a skew–symmetric bilinear function from $\mathbb{R}^{n_1} \times \mathbb{R}^{n_1}$ to $\mathbb{R}^{n_2}$ with integer coefficients when expressed in the standard bases of $\mathbb{R}^{n_1}$ and $\mathbb{R}^{n_2}$. It is easy to verify that G is a stratified 2–step nilpotent Lie group with dilations

$$(2.2) \qquad \delta_t(x, y) = (tx, t^2 y).$$

We fix a positive integer k, and let

$$\Gamma = \{(a, b) : a_j \in \mathbb{Z}, \ j = 1, \ldots, n_1, \ kb_j \in \mathbb{Z}, \ j = 1, \ldots, n_1\}.$$

It is easy to verify that Γ is a discrete co–compact subgroup.

Let $T = \{x \in \mathbb{R}^{n_1} : 0 \le x_j < 1, j = 1, \ldots, n\}$ denote the standard tile in $\mathbb{R}^{n_1}$ and $\mathbb{R}^{n_1} = \cup_{a \in \mathbb{Z}^{n_1}} (T + a)$ denote the standard tiling of $\mathbb{R}^{n_1}$. Let $A_1, \ldots, A_{n_2}$ be bounded measurable real–valued functions on T, and denote by A the tile

$$(2.3) \qquad A = \left\{ (x, y) \in G : x \in T, 0 \le y_j - A_j(x) < \frac{1}{k}, j = 1, \ldots, n_2 \right\}.$$

If $\gamma = (a, b) \in \Gamma$ then the image γA of A under left translation by γ is

$$(2.4)$$
$$\gamma A =$$
$$\left\{ (x, y) : x - a \in T, 0 \le y_j - b_j - S_j(a, x - a) - A_j(x - a) < \frac{1}{k}, j = 1, \ldots, n_2 \right\}$$

and $G = \cup_{\gamma \in \Gamma} \gamma A$ is a tiling of G. We call such a tiling a <u>stacked tiling</u> from the obvious geometric picture of a stack of tiles lying over each tile $T + a$. A stacked tiling is called <u>self–similar</u> if there exists a finite subset Γ_0 of Γ such that

$$(2.5) \qquad \delta_2 A = \bigcup_{\gamma \in \Gamma_0} \gamma A \quad \text{(disjoint union)}$$

or equivalently

$$(2.5') \qquad A = \bigcup_{\gamma \in \Gamma_0} \delta_{1/2} \gamma A \quad \text{(disjoint union)}.$$

To be specific we will require

$$(2.6) \qquad \Gamma_0 = \left\{ (a, b) : a_j = 0 \text{ or } 1, b_j = 0, \frac{1}{k}, \frac{2}{k}, \text{ or } \frac{3}{k} \right\},$$

but other choices are possible.

LEMMA 2.1. $G = \cup_{\gamma \in \Gamma} \gamma A$ is a self–similar stacked tiling with Γ_0 given by (2.6) if and only if the functions A_j on T satisfy

$$(2.7) \qquad A_j(x) = \frac{1}{4} A_j(\langle 2x \rangle) + \frac{1}{4} S_j([2x], \langle 2x \rangle)$$

where [ ] and ⟨ ⟩ denote the integer and fractional part functions, interpreted componentwise ([x]ⱼ = [xⱼ], etc.).

PROOF. Note that if $x - a \in T$ then $a = [x]$ and $x - a = \langle x \rangle$. Then an easy calculation using (2.4) shows

$$\bigcup_{\gamma \in \Gamma_0} \gamma A = \Big\{(x, y) : 0 \leq x_j < 2, j = 1, \ldots, n_1, \quad \text{and}$$

$$0 \leq y_j - S_j([x], \langle x \rangle) - A_j(\langle x \rangle) < \frac{4}{k}, j = 1, \ldots, n_2\Big\},$$

while

$$\delta_2 A = \Big\{(x, y) : 0 \leq x_j < 2, j = 1, \ldots, n_1, \quad \text{and}$$

$$0 \leq \frac{1}{4} y_j - A_j(\frac{1}{2} x) < \frac{1}{k}, j = 1, \ldots, n_2\Big\}.$$

For these two sets to be equal we must have

$$A_j(\frac{1}{2} x) = \frac{1}{4} A_j(\langle x \rangle) + \frac{1}{4} S_j([x], \langle x \rangle)$$

for all x satisfying $0 \leq x_j < 2$, and replacing x_j by $2x_j$ yields (2.7).   Q.E.D.

THEOREM 2.2. *There exists a unique self–similar stacked tiling $G = \bigcup_{\gamma \in \Gamma} \gamma A$ with Γ_0 given by (2.6). The functions $A_j(x)$ are given explicitly by*

$$(2.8) \qquad A_j(x) = \sum_{n=1}^{\infty} \frac{1}{4^n} S_j([2^n x] \bmod 2, \langle 2^n x \rangle)$$

where $[2^n x_j] \bmod 2$ means 0 or 1 depending on the parity $[2^n x_j]$.

PROOF. The mapping $MA_j(x) = \frac{1}{4} A_j(\langle 2x \rangle) + \frac{1}{4} S_j([2x], \langle 2x \rangle)$ is a contractive affine mapping of $L^\infty(T)$, so it has a unique fixed point, given by $\lim_{n \to \infty} M^n 0$ which yields (2.8).   Q.E.D.

COROLLARY 2.3. *The tile A has an interior of full Lebesgue measure (k^{-n_2}).*

PROOF. Let T_0 denote the subset of T of all x such that no x_j is a dyadic rational (integer times 2^{-n}). Although the functions A_j are in general discontinuous, having jump discontinuities at the points not in T_0, it is easy to see that they are continuous at all points of T_0. Thus the points (x, y) for $x \in T_0$ and $0 < y_j - A_j(x) < \frac{1}{k}$ lie in the interior of A, and this set has full measure. Q.E.D.

The boundary ∂A of A splits naturally into $2(n_1 + n_2)$ faces of two types. There are smooth faces given by an identity $x_j = 0$ (or 1), and fractal faces given by $y_j = A_j(x)$ (or $A_j(x) + \frac{1}{k}$). We will compute the dimensions of these faces. First we recall some basic facts about metrics and dimension on G. Let d denote any metric on G that is invariant under left translations and homogeneous of degree one under dilation. It is known that such metrics exist, and any two are equivalent. Given s uch a metric we can define Hausdorff measure of dimension α

on G for every real α. The Hausdorff dimension of G is known to be $n_1 + 2n_2$, and Hausdorff measure of dimension $n_1 + 2n_2$ is a positive multiple of Haar measure (the determination of the constant is an open problem — it depends, of course, on the choice of metric). We can also define upper and lower Minkowski content of dimension α, and Minkowski dimension for subsets of G, with $\dim_M \geq \dim_H$. Again $\dim_M(G) = n_1 + 2n_2$.

Let $F \subseteq \partial A$ be the fractal face defined by the condition $y_1 = A_1(x)$. Let $\Gamma_1 \subseteq \Gamma_0$ be defined by the condition $b_1 = 0$. Note that the cardinality of Γ_1 is $2^{n_1 + 2n_2 - 2}$, one quarter that of Γ_0. It follows easily that

$$(2.9) \qquad F = \bigcup_{\gamma \in \Gamma_1} \delta_{1/2}\gamma F$$

and this is the key observation for showing that the dimension of F is $n_1 + 2n_2 - 2$.

THEOREM 2.4. $\dim_H(F) = \dim_M(F) = n_1 + 2n_2 - 2$ and the $n_1 + 2n_2 - 2$ dimensional Hausdorff measure of F is finite and positive.

PROOF. By iterating (2.9), it follows that there exists a subset Γ_m of Γ of cardinality $\leq 2^{(n_1 + 2n_2 - 2)m}$ such that

$$(2.10) \qquad F = \bigcup_{\gamma \in \Gamma_m} \delta_{1/2^m}\gamma F.$$

Thus we have covered F by at most $2^{(n_2 + 2n_2 - 2)m}$ sets of diameter exactly $c2^{-m}$ (c is the diameter of F). This shows $\dim_M(F) \leq n_1 + 2n_2 - 2$ and that the $n_1 + 2n_2 - 2$ dimensional Hausdorff measure of F is finite. To complete the proof we need to show that this measure is positive.

Suppose we have a covering $F \subseteq \cup_{j=1}^{\infty} V_j$ by sets of diameter $d_j \leq \varepsilon$. We need a lower bound for $\sum_{j=1}^{\infty} d_j^{n_1 + 2n_2 - 2}$. Without loss of generality we may assume $1/kd_j^2$ is an integer. We construct a covering of A by associating to each V_j, $1/kd_j^2$ other sets $V_{j,r}$ for $r = 1, 2, \ldots, 1/kd_j^2$, where $V_{j,r} = V_j + I_r$ and I_r is just the interval along the y_1 axis from $(r-1)d_j^2$ to rd_j^2. It is clear from the definition of A and F that A is covered by the sets $V_{j,r}$. Also, the interval I_r is of length d_j^2 so in the metric it has diameter on the order of d_j, so we have $\operatorname{diam}(V_{j,r}) \leq cd_j$ for a fixed constant c. Thus we have

$$(2.11) \qquad \sum_{j,r} \operatorname{diam}(V_{j,r})^{n_1 + 2n_2} \leq c' \sum d_j^{n_1 + 2n_2 - 2}.$$

But if we take ε small enough the left side of (2.11) is bounded away from zero because A has positive $n_1 + 2n_2$ dimensional Hausdorff measure, since A contains an open set. This gives the lower bound for the right side of (2.11), and so proves that the $n_1 + 2n_2 - 2$ dimensional Hausdorff measure of F is positive. Q.E.D.

This conclusion is surprising, and helps explain the fractal nature of the face. The smooth faces of ∂A have dimension $n_1 + 2n_2 - 1$, and this will be true of any smooth hypersurface in G. It is easiest to present the argument in the lowest dimensional case, $n_1 = 2$, $n_2 = 1$. Let X_1, X_2 and Y_1 denote the left–invariant vectors associated to the coordinate directions, so Y_1 is in the center of the Lie algebra and $[X_1, X_2] = cY_1$. A curve in G is said to be <u>horizontal</u> if its tangent vector at every point lies in the two–dimensional space spanned by X_1 and X_2. For horizontal curves, the arclength along the curve is equivalent to the metric distance, so Hausdorff dimension is one and the Hausdorff measure of dimension one is comparable to arclength measure. However, for any non–horizontal curve (this is the generic case), the metric distance is considerably longer than arclength, in fact it is comparable to the square root of arclength. This makes the curve have Hausdorff dimension two.

Now consider any smooth surface. It is impossible that all curves in the surface are horizontal, for that would violate the Frobenius theorem since X_1 and X_2 do not span a Lie subalgebra. But the tangent space to the surface must intersect this two dimensional subspace in at least a one dimensional subspace. Thus we can find a piece of the surface that is a product of a one–dimensional horizontal curve and a two-dimensional non–horizontal curve. Thus the surface has Hausdorff dimension $3 = n_1 + 2n_2 - 1$. See [Mi], [NSW], [S6] and [S7] for further discussion of the geometry here.

§3. Tilings of Rational Graded Groups

In this section we consider the most general groups possessing both dilations and lattice subgroups, what we call <u>rational graded nilpotent Lie groups of step r</u>. We give the definition in terms of the Lie group rather than the Lie algebra, but the reader can easily translate if this is desired. Here $G = \mathbb{R}^{n_1} \times \mathbb{R}^{n_2} \times \cdots \times \mathbb{R}^{n_r}$ with elements x in G written $x = (x_1, \ldots, x_r)$ for $x_i \in \mathbb{R}^{n_i}$, or $x = x_{ij}$ for $1 \leq i \leq r$, $1 \leq j \leq n_i$, and group law

$$(3.1) \qquad (x \circ x')_{ij} = x_{ij} + x'_{ij} + F_{ij}(x, x')$$

where F_{ij} is a polynomial in $x_1, \ldots, x_{i-1}, x'_1, \ldots, x'_{i-1}$ with rational coefficients which satisfies the homogeneity condition

$$(3.2) \qquad F_{ij}(\delta_t x, \delta_t x') = t^i F_{ij}(x, x')$$

for the dilations

$$(3.3) \qquad (\delta_t x)_{ij} = t^i x_{ij}.$$

(In particular, $F_{1j} \equiv 0$.) The homogeneity condition i mplies that the dilations act as automorphisms on G. (The requirement that G be a group imposes rather stringent conditions on the functions F_{ij} that we will not make explicit.)

Let $k_1, k_2, \ldots, k_r$ be positive integers and define $\Gamma \subseteq G$ by $a = a_{ij} \in \Gamma$ if and only if $k_i a_{ij} \in \mathbb{Z}$. The condition that the coefficients of the polynomials F_{ij} be rational easily implies that it is possible to choose the integers $k_1, \ldots, k_r$ in such a way that Γ is a subgroup of G. We assume that such a choise is made, and the integers $k_1, \ldots, k_r$ will be fixed throughout the remainder of the paper. Of course Γ is cocompact.

A tile is a measurable subset T of G such that

$$(3.4) \qquad G = \bigcup_{\gamma \in \Gamma} \gamma^{-1} T \quad \text{(disjoint union)}$$

and the decomposition (3.4) is called a tiling. A tiling is said to be self–similar if there exists a finite subset Γ_0 of Γ such that

$$(3.5) \qquad \delta_2 T = \bigcup_{\gamma \in \Gamma_0} \gamma^{-1} T \quad \text{(disjoint union)}$$

or equivalently

$$(3.5') \qquad T = \bigcup_{\gamma \in \Gamma_0} \delta_{1/2} \gamma^{-1} T \quad \text{(disjoint union)}.$$

We will restrict attention to the case

$$(3.6) \qquad \Gamma_0 = \{a_{ij} : k_i a_{ij} = 0, -1, -2, \ldots, -(2^i - 1)\}$$

although other choices are possible. We will also restrict attention to tiles given by

$$(3.7) \qquad T = \left\{ x : 0 \leq x_{ij} - A_{ij}(x) < \frac{1}{k_i} \right\}$$

where A_{ij} are bounded measurable functions of $x_1, \ldots, x_{i-1}$ (in particular $A_{1j} \equiv 0$). It is easy to see that any choice of the functions A_{ij} yields a tile. We call the associated tiling a stacked tiling.

LEMMA 3.1. *Any T given by (3.7) is a tile, and $x \in \gamma^{-1} T$ for $\gamma = a$ defined recursively by*

$$(3.8) \qquad a_{ij} = -k_i^{-1}[k_i x_{ij} + k_i F_{ij}(a, x) - k_i A_{ij}(a + x + F(a, x))]$$

(here [ ] denotes greatest integer part).

PROOF. $\gamma x \in T$ if and only if $0 \leq a_{ij} + x_{ij} + F_{ij}(a, x) - A_{ij}(a + x + F(a, x)) < \frac{1}{k_i}$ by (3.1) and (3.7). But since $k_i a_{ij}$ is an integer this is equivalent to (3.8). Q.E.D.

LEMMA 3.2. *The stacked tiling defined by (3.7) is self–similar if and only if the functions A_{ij} satisfy*

$$(3.9) \qquad A_{ij}(a(x) + x + F(a(x), x)) = F_{ij}(a(x), x) + 2^i A_{ij}(\delta_{1/2} x)$$

where $a(x)$ is defined by (3.8).

PROOF. $x \in \delta_2 T$ if and only if $\delta_{1/2} x \in T$ which means

$$(3.10) \qquad 0 \leq k_i x_{ij} - 2^i k_i A_{ij}(\delta_{1/2} x) < 2^i.$$

On the other hand, $x \in \cup_{\gamma \in \Gamma_0} \gamma^{-1}T$ if and only if there exists a_{ij} with $0 \le -k_i a_{ij} < 2^i$ such that (3.8) holds; in other words

(3.11) $0 \le k_i x_{ij} + k_i F_{ij}(a, x) - k_i A_{ij}(a + x + F(a, x)) < 2^i.$

For (3.10) and (3.11) to be equivalent it is necessary and sufficient that (3.9) holds.

                                                                Q.E.D.

THEOREM 3.3. *There exists a unique self–similar stacked tiling, with T given by (3.7) and $A_{ij}(x)$ given explicitly by*

(3.12) $A_{ij}(x) = - \sum_{n=1}^{\infty} 2^{-in} F_{ij}(a(\delta_2(R^{n-1}x), \delta_2 R^{n-1}x)$

where $a(x)$ is given by (3.8) and

(3.13) $Rx = a(\delta_2 x) + \delta_2 x + F(a(\delta_2 x), \delta_2 x) = a(\delta_2 x) \circ \delta_2 x.$

Furthermore, the interior of T has full Lebesgue measure

PROOF. By Lemma 3.2 it suffices to solve simultaneously (3.9) and (3.8). We rewrite (3.9) as

(3.14) $A_{ij}(x) = 2^{-i} A_{ij}(Rx) - 2^{-i} F_{ij}(a(\delta_2 x), \delta_2 x).$

We solve (3.14) and (3.2) inductively on i. We already know $A_{1j}(x) \equiv 0$, and then (3.8) gives

$$a_{1j}(x) = -k_1^{-1}[k_1 x_{1j}].$$

For the induction step, assume $A_{i'j}$ and $a_{i'j}$ are solutions of (3.14) and (3.8) for all $i' < i$. Since $F_{ij}(a(\delta_2 x), \delta_2 x)$ depends only on $a_{i'j}(\delta_2 x)$ for $i' < i$, equation (3.14) is a fixed–point equation for the mapping on L^∞ defined explicitly by the right side of (3.14). Since this is clearly an affine contraction, the unique solution is given by (3.12). Then (3.8) gives a_{ij} in terms of previously determined quantities. Finally, the proof that the interior of T has full Lebesgue measure is essentially the same as the proof of Corollary 2.3.   Q.E.D.

§4. Maximal Functions and Hausdorff Measures

In this section we assume that G is a rational graded nilpotent Lie group of step r. We choose a metric $d(x, y)$ on G which satisfies

(4.1) $d(x, y) = d(zx, zy)$ for all $x, y, z \in G$ (left translation invariance)

(4.2) $d(\delta_t x, \delta_t y) = td(x, y)$ for all $x, y \in G, t > 0$ (dilation homogeneity)

(4.3) d induces the usual topology on G.

It is well–known that such metrics exist, and any two are equivalent ([FS], [NSW]). We let $B_r(x)$ denote the open ball of radius r about the point x, with respect to the metric.

Let μ be any locally finite positive measure on the Borel sets of G. We introduce a variety of maximal functions associated with μ:

DEFINITION 4.1. The <u>centered maximal function</u> M_μ is

$$(4.4) \qquad M_\mu f(x) = \sup_{r>0} \frac{1}{\mu(B_r(x))} \int_{B_r(x)} |f| d\mu$$

(strictly speaking, the supremum is taken over all r for which $\mu(B_r(x)) > 0$). The <u>local centered maximal function</u> m_μ is

$$(4.5) \qquad m_\mu f(x) = \sup_{0<r\le 1} \frac{1}{\mu(B_r(x))} \int_{B_r(x)} |f| d\mu.$$

The <u>α–dimensional maximal function</u> M_α is

$$(4.6) \qquad M_\alpha f(x) = \sup_{r>0} r^{-\alpha} \int_{B_r(x)} |f| d\mu$$

and the local version m_α is

$$(4.7) \qquad m_\alpha f(x) = \sup_{0<r\le 1} r^{-\alpha} \int_{B_r(x)} |f| d\mu.$$

The <u>limsup α–dimensional maximal function</u> $\overline{M}_\alpha$ is

$$(4.8) \qquad \overline{M}_\alpha f(x) = \limsup_{r\to 0} r^{-\alpha} \int_{B_r(x)} |f| d\mu.$$

The <u>tiling maximal function</u> M_T is

$$(4.9) \qquad M_T f(x) = \sup_{j\in\mathbb{Z}} \frac{1}{\mu(T_j(x))} \int_{T_j(x)} |f| d\mu$$

where $T_j(x)$ denotes the dilated tile $\delta_{2^{-j}} \gamma^{-1} T$ which contains x (for each j there is exactly one $\gamma \in \Gamma$ for which this is true), and its local version m_T is

$$(4.10) \qquad m_T f(x) = \sup_{j\ge 0} \frac{1}{\mu(T_j(x))} \int_{T_j(x)} |f| d\mu.$$

Next we introduce various conditions on the measure μ that will enter as hypotheses in many of our results:

DEFINITION 4.2. μ satisfies the <u>doubling condition</u> if there exists a constant c such that

$$(4.11) \qquad \mu(B_{2r}(x)) \le c\mu(B_r(x))$$

for μ–almost every x. μ is said to be <u>uniformly α–dimensional</u> if there exists a constant c such that

$$(4.12) \qquad \mu(B_r(x)) \le cr^\alpha \text{ for all } x \in G$$

and all $r > 0$, and <u>locally uniformly α–dimensional</u> if (4.12) holds for all $r \leq 1$. We say that μ is <u>α–dimensional in the L^p sense</u> if there exists a constant c such that

$$\text{(4.13)} \qquad \sum_{\gamma \in \Gamma} |\mu(\delta_{2^{-j}} \gamma^{-1} T)|^p \leq c 2^{-j\alpha(p-1)}$$

for all $j \geq 0$. We denote by $\mathcal{M}_\alpha^p$ the space of all such measures, and we let $\|\mu\|_{\mathcal{M}_\alpha^p}^p$ be the smallest constant in (4.13).

Let Q denote the homogeneous dimension of the group

$$\text{(4.14)} \qquad Q = \sum_{i=1}^r i n_i$$

and let dx denote Lebesgue measure on G, which is homogeneous of degree Q with respect to the dilations δ_t.

LEMMA 4.3. *μ is α–dimensional in the L^p sense if and only if*

$$\text{(4.15)} \qquad \int_G |\mu(B_r(x))|^p dx \leq c r^{Q+\alpha(p-1)}$$

for all $r \leq 1$, with $c \approx \|\mu\|_{\mathcal{M}_\alpha^p}^p$.

We omit the proof, which is a routine consequence of the fact that tiles fit inside balls and balls fit inside unions (uniformly bounded in number) of tiles, with comparable sizes (see [L]).

Now it is known that G has a covering lemma of Wiener type ([KV]), but not of Besicovitch type ([W], [KR]). This means that the centered maximal function M_μ is of weak–type $(1,1)$ and is bounded on L^p, $1 < p \leq \infty$, with respect to the measure μ, provided μ satisfies the doubling condition, but we cannot expect these estimates to hold for M_μ without this assumption. On the other hand, since the dyadic tiles $\delta_{2^{-j}} \gamma^{-1} T$ form a net of sets (if two intersect then one contains the other), it is straightforward to prove the desired estimates for the tiling maximal function M_T for any measure μ. Unfortunately, the tiling maximal function is not large enough to control the kind of averaging processes we will consider in the next section.

Since G is a metric space with respect to the distance function d, we can define Hausdorff measures of all dimensions, and we denote the α–dimensional Hausdorff measure by μ_α. It is known that G has dimension Q, and $\mu_Q = c dx$ for a positive constant c ([M]). The constant depends on the choice of the metric, and may be difficult to compute exactly because the isodiametric problem is not as simple as in the Euclidean case. This problem, to find the set of fixed diameter of maximal measure, is solved by a ball in $\mathbb{R}^n$, as can be seen using Steiner symmetrization, for example ([E]). However, for some choices of metric on G, balls are not convex, hence cannot solve the problem.

An important tool in the study of Hausdorff measure is a Vitali–type covering lemma, which is valid in any metric space.

LEMMA 4.4. *Let E be a Borel set, and let $\mathcal{B}$ be a collection of closed balls such that for each $x \in E$ and $\varepsilon > 0$ there exists $r \leq \varepsilon$ with $B_r(x) \in \mathcal{B}$. Then there is a finite or countable disjoint sequence $\{B_{r_j}(x_j)\}$ of balls in $\mathcal{B}$ such that either*

(4.16) $$\sum r_j^\alpha = +\infty$$

or

(4.17) $$\mu_\alpha\left(E \setminus \bigcup_j B_{r_j}(x_j)\right) = 0.$$

Furthermore, if $\mu_\alpha(E) < \infty$ then for every $\varepsilon > 0$ we can choose the sequence so that

(4.18) $$\mu_\alpha(E) \leq \sum (2r_j)^\alpha + \varepsilon.$$

PROOF. The proof is the same as in Euclidean space ([F1], Theorem 1.10).
$$\text{Q.E.D.}$$

Using this covering lemma, we can prove the basic facts about α–dimensional densities, just as in the Euclidean case. If E is a Borel set define the upper and lower α–dimensional densities as follows:

(4.19) $$\overline{D}^\alpha(E, x) = \limsup_{r \to 0} (2r)^{-\alpha} \mu_\alpha(E \cap B_r(x))$$

(4.20) $$\underline{D}_\alpha(E, x) = \liminf_{r \to 0} (2r)^{-\alpha} \mu_\alpha(E \cap B_r(x))$$

We say E is <u>regular</u> if

$$\overline{D}^\alpha(E, x) = \underline{D}_\alpha(E, x) = 1 \text{ for } \mu_\alpha - a.e. \; x \in E.$$

We say E is <u>quasi–regular</u> if there exists $\varepsilon > 0$ such that

$$\underline{D}_\alpha(E, x) \geq \varepsilon \text{ for } \mu_\alpha - a.e. \; x \in E.$$

THEOREM 4.5. *Suppose $\mu_\alpha(E) < \infty$. Then*

(4.21) $$\overline{D}^\alpha(E, x) = 0 \text{ for } \mu_\alpha - \text{ almost every } x \text{ not in } E,$$

and

(4.22) $$2^{-\alpha} \leq \overline{D}^\alpha(E, x) \leq 1 \text{ for } \mu_\alpha - \text{ almost every } x \text{ in } E.$$

PROOF. See ([F1], Corollary 2.4, 2.5).
$$\text{Q.E.D.}$$

We can also use the covering lemma to control the limsup maximal function if μ is μ_α restricted to a set E, and μ is locally uniformly α–dimensional.

THEOREM 4.6. *Let $\mu = \mu_\alpha|_E$ for some Borel set E, and assume μ is also locally uniformly α–dimensional. Then $\overline{M}_\alpha$ is weak–type $(1,1)$ and bounded on L^p, $1 < p \leq \infty$, with respect to the measure μ.*

PROOF. It suffices to show

$$(4.23) \qquad \mu\{x : \overline{M}_\alpha f(x) > s\} \leq c\|f\|_{L^1(d\mu)}/s, \text{ for } f \in L^1(d\mu).$$

Assume first $\mu_\alpha(E) < \infty$. Let $F_s = \{x \in E : \overline{M}_\alpha f(x) > s\}$. By definition, for each $x \in F_s$ and $\varepsilon > 0$ there exists $r \leq \varepsilon$ such that

$$(4.24) \qquad \int_{B_r(x) \cap E} |f| d\mu_\alpha \geq s r^\alpha.$$

Let $\mathcal{B}$ be the collection of such balls, and apply Lemma 4.4 to obtain a disjoint sequence $\{B_{r_j}(x_j)\}$. From (4.24) we have

$$(4.25) \qquad \int_{E \cap \cup B_{r_j}(x_j)} |f| d\mu_\alpha \geq s \sum r_j^\alpha$$

which means (4.16) is impossible, so (4.17) must hold. But this means

$$\mu_\alpha(F_s) = \mu_\alpha(F_s \cap \cup B_{r_j}(x_j)) = \sum \mu_\alpha(F_s \cap B_{r_j}(x_j)).$$

Now since $\mu_\alpha|_E$ is locally uniformly α–dimensional we have

$$\mu_\alpha(F_s \cap B_{r_j}(x_j)) \leq \mu_\alpha(E \cap B_{r_j}(x_j)) \leq c r_j^\alpha$$

hence

$$\mu(F_s) \leq c \sum r_j^\alpha \leq c\|f\|_{L^1(d\mu)}/s$$

by (4.25).

If $\mu_\alpha(E) = +\infty$ we simply apply the above argument to $E \cap \gamma^{-1}T$ and then sum over $\gamma \in \Gamma$. Q.E.D.

COROLLARY 4.7. *Let f be a non–negative function in $L^p(d\mu)$ for some p, $1 \leq p \leq \infty$, where μ is as in Theorem 4.6. Then*

$$(4.26) \qquad \limsup_{r \to 0} \frac{1}{(2r)^\alpha} \int_{B_r(x)} f d\mu \leq f(x) \text{ for } \mu_\alpha - a.e. \ x \in E,$$

$$(4.27) \qquad \lim_{r \to 0} \frac{1}{(2r)^\alpha} \int_{B_r(x)} f d\mu = 0 \text{ for } \mu_\alpha - a.e. \ x \notin E,$$

and

$$(4.28) \qquad \liminf_{r \to 0} \frac{1}{(2r)^\alpha} \int_{B_r(x)} f d\mu \geq \underline{D}_\alpha(E, x) f(x) \text{ for } \mu_\alpha - a.e. \ x \in E.$$

In particular, if E is regular then

$$(4.29) \qquad \lim_{r \to 0} \frac{1}{(2r)^\alpha} \int_{B_r(x)} f d\mu = f(x) \text{ for } \mu_\alpha - a.e. \ x \in E,$$

while if E is only quasi–regular then there exists $\varepsilon > 0$ such that

(4.30) $$\liminf_{r\to 0} \frac{1}{(2r)^\alpha} \int_{B_r(x)} f\,d\mu \geq \varepsilon f(x) \text{ for } \mu_\alpha - a.e.\ x \in E.$$

PROOF. Since μ is a regular Borel measure we can write $f = f_k + g_k$ where f_k is continuous and $\|g_k\|_p \leq 2^{-k}$. Then $\overline{M}_\alpha f(x) \leq \overline{M}_\alpha f_k(x) + \overline{M}_\alpha g_k(x)$. Since f_k is con tinuous

$$\overline{M}_\alpha f_k(x) \leq f_k(x) \overline{D}^\alpha(E, x)$$

for every x in E, so

(4.31) $$\overline{M}_\alpha f(x) \leq f_k(x) + \overline{M}_\alpha g_k(x) \text{ for } \mu_\alpha - a.e.\ x \in E,$$

by (4.22), and (4.26) follows from (4.31) and Theorem 4.8 by standard functional analysis arguments. A similar argument establishes (4.28), and then (4.29) and (4.30) are immediate consequences of the definitions.

We prove (4.27) by contradiction. Assume there exists a set F disjoint from E with $\mu_\alpha(F) > 0$ for which (4.27) fails. By the inner regularity of μ_α we may assume without loss of generality that F is compact. By a theorem of Besicovitch, F contains a compact subset F_0 with $\mu_\alpha(F_0) > 0$ such that F_0 is locally uniformly α–dimensional. (See [F1], Theorem 5.4 (b) for a proof that extends to our context, with the tilings of §3 playing the role of the dyadic cubes in Euclidean space. Although the hypothesis in that theorem is $\mu_\alpha(F) = \infty$, it is in fact not used in the proof of part (b).)

Next consider the measure $\tilde{\mu} = \mu_\alpha|_{E \cup F_0}$. It satisfies the hypotheses of Theorem 4.8. Let $\tilde{f}$ be the extension of f to $E \cup F_0$ by setting $\tilde{f}(x) = 0$ for $x \in F_0$. Now apply (4.26) to $\tilde{f}$ and $\tilde{\mu}$. We obtain

$$\limsup_{r\to 0} \frac{1}{(2r)^\alpha} \int_{B_r(x)} f\,d\mu = \limsup_{r\to 0} \frac{1}{(2r)^\alpha} \int_{B_r(x)} \tilde{f}\,d\tilde{\mu} \leq f(x) \text{ for } \mu_\alpha - a.e.\ x \in F_0,$$

which contradicts the assumption that (4.27) fails on F_0.   Q.E.D.

Given two measure μ and ν, we say ν is <u>null with respect to</u> μ, written $\nu \lll \mu$, if $\mu(E) < \infty$ implies $\nu(E) = 0$.

THEOREM 4.8. *Suppose ν is a locally finite measure with $\nu \lll \mu_\alpha$. Then*

(4.32) $$\lim_{r\to 0} \nu(B_r(x))/(2r)^\alpha = 0 \text{ for } \mu_\alpha - a.e.\ x.$$

If $\mu = \mu_\alpha|_E + \nu$ with E as in Theorem 4.6 and $\nu \lll \mu_\alpha$ then the conclusions of Corollary 4.7 continue to hold.

PROOF. We use a covering lemma of Wiener type ([FS], p. 53). Without loss of generality we may assume ν is finite and has bounded support. Let

$$E = \left\{ x \in G : \limsup_{r\to 0} (2r)^{-\alpha} \nu(B_r(x)) \geq 1/k \right\}.$$

It suffices to prove $\mu_\alpha(E) = 0$ for each k to establish (4.32).

Given any $\varepsilon > 0$, to each $x \in E$ there exists $r \leq \varepsilon$ such that

$$(4.33) \qquad\qquad \nu(B_r(x)) \geq (2r)^\alpha/k.$$

By the Wiener type covering lemma there exists a sequence $\{x_j\}$ in E (with corresponding $r_j \leq \varepsilon$) such that the balls $B_{r_j}(x_j)$ are disjoint and the balls $B_{\lambda r_j}(x_j)$ cover E (here λ is a constant that depends only on the group G). Let $\mu_{\alpha,\varepsilon}$ denote the ε–approximate outer measure to μ_α. Then

$$\mu_{\alpha,\varepsilon}(E) \leq \sum (2\lambda r_j)^\alpha \leq \lambda^\alpha k \sum \nu(B_{r_j}(x_j)) \leq \lambda^\alpha k\nu(E_{2\lambda\varepsilon})$$

by (4.33), where $E_{2\lambda\varepsilon}$ denotes the $2\lambda\varepsilon$ neighborhood of E. We let $\varepsilon \to 0$ and use the continuity from above for the finite measure ν to obtain $\mu_\alpha(E) \leq \lambda^\alpha k\nu(E)$. Since $\nu(E) < \infty$ this implies $\mu_\alpha(E) < \infty$ so $\nu(E) = 0$ because $\nu \lll \mu_\alpha$ hence finally $\mu_\alpha(E) = 0$.

If $f \in L^p(d\mu)$ for $\mu = \mu_\alpha|_E + \nu$ then $fd\nu \lll \mu_\alpha$ so by (4.32) it does not contribute to any of the limits in the conclusions of Corollary 4.7.   Q.E.D.

§5. Asymptotics of Heat Expansions

Now we choose a sub–Laplacian $\mathcal{L}$ on G,

$$(5.1) \qquad\qquad \mathcal{L} = \sum_{j=1}^{n_1} X_j^2$$

where $X_1, \ldots, X_{n_1}$ is a basis for the subspace of the Lie algebra of G generated by the first n_1 coordinates. We assume that the group is <u>stratified</u>, meaning that $X_1, \ldots, X_{n_1}$ generate the full Lie algebra $\mathfrak{g}$ of G. Then $\mathcal{L}$ generates a heat semigroup which we denote $e^{t\mathcal{L}}$ which is given by a heat kernel $h_t(x)$,

$$(5.2) \qquad\qquad e^{t\mathcal{L}}(\mu)(x) = \int_G h_t(y^{-1}x)d\mu(y)$$

where $h_t \in \mathcal{S}(G)$ and satisfies the homogeneity property

$$(5.3) \qquad\qquad h_t(x) = t^{-Q/2}h_1(\delta_{t^{-1/2}}x)$$

([FS]).

In this section our goal is to relate the asymptotic behavior of $e^{t\mathcal{L}}(\mu)$ as $t \to 0$ to the fractal properties of the measure μ.

THEOREM 5.1. *Suppose $|\mu| \in \mathcal{M}_\alpha^p$. Then*

$$(5.4) \qquad\qquad \sup_{t\leq 1} t^{(Q-\alpha)/2p'}\|e^{t\mathcal{L}}\mu\|_p \leq c\|\mu\|_{\mathcal{M}_\alpha^p},$$

(the L^p norm is with respect to Haar measure). Conversely, if μ is a positive measure then the inequality can be reversed.

PROOF. By Lemma 4.3 we have

$$(5.5) \qquad\qquad \|\,|\mu|(B_r(x))\|_p \leq c\|\mu\|_{\mathcal{M}_\alpha^p} r^{Q/p+\alpha/p'}$$

for $r \leq 1$. On the other hand, from (5.5) with $r = 1$ we obtain

$$(5.6) \qquad \| \, |\mu|(B_r(x))\|_p \leq c\|\mu\|_{\mathcal{M}_\alpha^p} r^Q$$

for $r \geq 1$. In fact it is easier to obtain the discrete analogue of (5.6), namely

$$(5.6') \qquad \sum_{\gamma \in \Gamma} |\mu|(\delta_{2^j}\gamma^{-1}T)^p \leq c\|\mu\|_{\mathcal{M}_\alpha^p} 2^{jQ(p-1)}.$$

For $j = 1$ we use (3.5) which is the same as

$$\delta_2\gamma^{-1}T = \bigcup_{\gamma_0 \in \Gamma_0} (\gamma_0\delta_2\gamma)^{-1}T \quad \text{(disjoint)}$$

and the cardinality of Γ_0 is 2^Q, so

$$|\mu|(\delta_2\gamma^{-1}T)^p = \left(\sum_{\gamma_0 \in \Gamma_0} |\mu|((\gamma_0\delta_2\gamma)^{-1}T) \right)^p \leq 2^{Q(p-1)} \left(\sum_{\gamma_0 \in \Gamma_0} |\mu|((\gamma_0\delta_2\gamma)^{-1}T)^p \right)$$

by Hölder's inequality. If we sum on γ (note that $\gamma_0\delta_2\gamma$ gives a parametrization of Γ as γ_0 varies in Γ_0 and γ varies in Γ) we obtain

$$\sum_{\gamma \in \Gamma} |\mu|(\delta_2\gamma^{-1}T)^p \leq 2^{Q(p-1)} \sum_{\gamma \in \Gamma} |\mu|(\gamma^{-1}T)^p$$

which gives (5.6') for $j = 1$. The general case follows by iteration, and (5.6') implies (5.6) by the same reasoning as in Lemma 4.3.

Since $h_1 \in \mathcal{S}$ we have the estimate

$$(5.7) \qquad |e^{t\mathcal{L}}\mu(x)| \leq c_N t^{-Q/2} \int (1 + t^{-1/2}d(x,y))^{-N}d|\mu|(y).$$

Now we can write

$$(5.8) \qquad \int (1 + t^{-1/2}d(x,y))^{-N}d|\mu|(y)$$

$$= Nt^{-1/2} \int_0^\infty |\mu|(B_r(x))(1 + t^{-1/2}r)^{-N-1}dr.$$

Substituting (5.8) in (5.7) and using Minkowski's inequality we obtain

$$(5.9) \qquad \|e^{t\mathcal{L}}\mu\|_p \leq ct^{-Q/2} \int_0^\infty \| \, |\mu|(B_r(x))\|_p t^{-1/2}(1 + t^{-1/2}r)^{-N-1}dr.$$

Substituting (5.5) and (5.6) in (5.9) we obtain

$$\|e^{t\mathcal{L}}\mu\|_p \leq c\|\mu\|_{\mathcal{M}_\alpha^p} t^{-Q/2} \left(\int_0^1 r^{Q/p+\alpha/p'} t^{-1/2}(1 + t^{-1/2}r)^{-N-1}dr \right.$$

$$(5.10) \qquad \left. + \int_1^\infty r^Q t^{-1/2}(1 + t^{-1/2}r)^{-N-1}dr \right).$$

If we choose N sufficiently large we obtain (5.4) by elementary estimates (the integral for $0 \leq r \leq 1$ can be extended to $0 \leq r < \infty$ and evaluated as $ct^{Q/2p+\alpha/2p'}$, while the integral for $1 \leq r < \infty$ can be estimated using $(1 + t^{-1/2}r)^{-N-1} \leq t^{N/4}(1 + r)^{-N/2-1}$).

The converse is easier. We observe that $h_t(0) = \int_G h_{t/2}(x)^2 dx > 0$ (in fact $h_t(x) > 0$ for all $x \in G$ but this requires a more sophisticated proof, $h_t(x) \geq 0$ is standard [FS]). Thus there exists constants ε and c such that $\chi_{B_\varepsilon} \leq c h_1$ hence

$$(5.11) \qquad \chi_{B_r} \leq c t^{Q/2} h_t \text{ for } r = \varepsilon t^{1/2}$$

by (5.3). If μ is positive then

$$(5.12) \qquad \mu(B_r(x)) \leq c r^Q e^{r^2 \mathcal{L}}(\mu)(x)$$

hence

$$(5.13) \qquad \|\mu(B_r)\|_p \leq c r^Q \|e^{r^2 \mathcal{L}} \mu\|_p$$

and the reverse of (5.4) follows by taking the supremum over $r \leq 1$.   Q.E.D.

COROLLARY 5.2. *Suppose μ is a locally uniformly α dimensional positive measure and $f \in L^p(d\mu)$. Then*

$$(5.14) \qquad \sup_{t \leq 1} t^{(Q-\alpha)/2p'} \|e^{t\mathcal{L}}(f d\mu)\|_p \leq c \|f\|_{L^p(d\mu)}.$$

PROOF. It follows from Hölder's inequality that $|f| d\mu \in \mathcal{M}_\alpha^p$ so (5.14) follows from (5.4).   Q.E.D.

When $\alpha = 0$ we easily obtain the analogue of Wiener's Theorem. Note that a measure is in $\mathcal{M}_0^p$ if and only if $\sum_{\gamma \in \Gamma} |\mu(\gamma^{-1} T)|^p < \infty$.

THEOREM 5.3. *Suppose $|\mu| \in \mathcal{M}_0^2$ and let $\mu = \mu_c + \sum c_j \delta_{a_j}$ be the decomposition into continuous and discrete parts. Then*

$$(5.15) \qquad \lim_{t \to 0} t^{Q/2} \|e^{t\mathcal{L}} \mu\|_2^2 = A \sum |c_j|^2$$

where $A = 2^{-Q/2} h_1(0)$.

PROOF. We have

$$t^{Q/2} \|e^{t\mathcal{L}} \mu\|_2^2 = t^{Q/2} \iint h_{2t}(y^{-1} x) d\mu(x) d\overline{\mu}(y)$$

$$= 2^{-Q/2} \iint h_1(\delta_{(2t)^{-1/2}}(y^{-1} x)) d\mu(x) d\overline{\mu}(y).$$

Because h_1 vanishes at infinity, the integrand tends pointwise as $t \to 0$ to $h_1(0)$ times the characteristic function of $x = y$, and so (5.15) will follow if we can interchange the limit and the integral. This will be justified by the dominated convergence theorem. Since $h_1 \in \mathcal{S}$ we have

$$h_1(\delta_{(2t)^{-1/2}} y^{-1} x) \leq c_N (1 + d(x,y))^{-N}$$

for all $t \leq 1$ and all N, so we need to verify that

$$\iint (1 + d(x,y))^{-N} d|\mu|(x) d|\mu|(y)$$

is finite for N sufficiently large. But the double integral is equivalent to the double sum

$$\sum_{\gamma_1 \in \Gamma} \sum_{\gamma_2 \in \Gamma} (1 + d(\gamma_1^{-1}T, \gamma_2^{-1}T))^{-N} |\mu|(\gamma_1^{-1}T) |\mu|(\gamma_2^{-1}T)$$

and by Cauchy–Schwartz it suffices to show that

$$\sum_{\gamma_1 \in \Gamma} \sum_{\gamma_2 \in \Gamma} (1 + d(\gamma_1^{-1}T, \gamma_2^{-1}T))^{-N} |\mu|(\gamma_1^{-1}T)^2$$

is finite. But this follows from the hypothesis $\mu \in \mathcal{M}_0^2$ and the uniform boundedness of

$$\sum_{\gamma_2 \in \Gamma} (1 + d(\gamma_1^{-1}T, \gamma_2^{-1}T))^{-N}$$

which holds for $N > Q$.   Q.E.D.

REMARK. In some cases we can evaluate the constant A explicitly. Suppose G is the $(2n+1)$–dimensional Heisenberg group and $\mathcal{L}$ is the usual sub–Laplacian (in the notation of §2 $n_1 = 2n$, $n_2 = 1$ and

$$S(x, x') = 2 \sum_{j=1}^{n} (x_{n+j}x'_j - x_j x'_{n+j}).$$

Then $Q = 2n + 2$ and

(5.16) $$h_1(0) = (4\pi)^{-n-1} \int_0^\infty \left(\frac{s}{\sinh s}\right)^n ds$$

(see [G]). In particular, when $n = 1$

$$\int_0^\infty \frac{s}{\sinh s} ds = \pi^2/4 \text{ so } A = 1/64$$

and when $n = 2$

$$\int_0^\infty \left(\frac{s}{\sinh s}\right)^2 ds = \pi^2/6 \text{ so } A = 1/384\pi.$$

It should be possible to evaluate the integral in (5.16) in the general case, but I have not been able to find it in standard tables of integrals.

In the next results we consider $e^{t\mathcal{L}}(f d\mu)$ for $f \in L^2(d\mu)$ where μ is locally uniformly α–dimensional, and $\mu = \mu_\alpha|_E + \nu$ with $\nu \ll \mu_\alpha$. (In general μ being locally uniformly α–dimensional implies $\mu \ll \mu_\alpha$ so $\mu = \varphi d\mu_\alpha + \nu$ [S1]. We are therefore assuming that φ is the characteristic function of a set E.)

THEOREM 5.4. Let $\mu = \mu_\alpha|_E + \nu$ with $\nu \ll \mu_\alpha$ be locally uniformly α–dimensional and let $f \in L^p(d\mu)$ for some p, $1 \leq p \leq \infty$. Then

(5.17) $$\lim_{t \to 0} t^{(Q-\alpha)/2} e^{t\mathcal{L}}(f d\mu)(x) = 0$$

for μ_α almost every x not in E, and

(5.18) $$\limsup_{t \to 0} t^{(Q-\alpha)/2} |e^{t\mathcal{L}}(f d\mu)(x)| \leq c|f(x)|$$

for μ_α almost every x in E. If E is quasi–regular then

$$(5.19) \qquad \liminf_{t \to 0} t^{(Q-\alpha)/2} |e^{t\mathcal{L}}(f d\mu)(x)| \geq c|f(x)|$$

for μ_α almost every x in E.

PROOF. By a variant of (5.7) and (5.8) we have the estimate

$$(5.20) \quad t^{(Q-\alpha)/2}|e^{t\mathcal{L}}(fd\mu)| \leq c_N t^{-(\alpha+1)/2} \int_0^\infty (1 + t^{-1/2}r)^{-N-1} \left| \int_{B_r(x)} f d\mu \right| dr.$$

Now we always have the crude estimate

$$(5.21) \qquad \left| \int_{B_r(x)} f d\mu \right| \leq \begin{cases} cr^{\alpha/p'} \|f\|_p & r \leq 1 \\ cr^{Q/p'} \|f\|_p & r > 1 \end{cases}$$

by Hölder's inequality and the local uniform α–dimensionality of μ. Now for μ_α almost every x, we can use (4.26), (4.27) and (4.32) to estimate $|\int_{B_r(x)} f d\mu|$ for $r \leq \varepsilon$ (the ε depends on x), and (5.21) for $r > \varepsilon$. Substituting this in (5.20) yields (5.17) and (5.18) by standard estimates.

Similarly, if f is non–negative then (4.28) implies (5.19) since $h_1(x) > 0$. For f real–valued write $f = f^+ - f^-$ and split E accordingly, $E = E^+ \cup E^-$. Then for $x \in E^+$ we can omit the contribution of $e^{t\mathcal{L}}(f^- d\mu)$ by (5.17). Finally, for f complex–valued consider the real and imaginary parts separately.   Q.E.D.

To obtain L^2 estimates we require the additional hypothesis that μ satisfy the doubling condition, in order to control the centered maximal function.

THEOREM 5.5. *Let μ be as in Theorem 5.4, and assume also that μ satisfies the doubling condition. Let $f \in L^2(d\mu)$. Then*

$$(5.22) \qquad \limsup_{t \to 0} t^{(Q-\alpha)/2} \|e^{t\mathcal{L}}(fd\mu)\|_2^2 \leq c \int_E |f|^2 d\mu_\alpha.$$

If, in addition, E is quasi–regular, then

$$(5.23) \qquad \liminf_{t \to 0} t^{(Q-\alpha)/2} \|e^{t\mathcal{L}}(fd\mu)\|_2^2 \geq c \int_E |f|^2 d\mu_\alpha.$$

PROOF. We have

$$(5.24) \qquad t^{(Q-\alpha)/2} \|e^{t\mathcal{L}}(fd\mu)\|_2^2 = t^{(Q-\alpha)/2} \int e^{2t\mathcal{L}}(fd\mu)(x)\overline{f(x)}d\mu(x).$$

If we can apply the dominated convergence theorem to the right side of (5.23), then Theorem 5.4 gives (5.22). But the integrand is easily dominated by a multiple of $M_\mu f(x)|f(x)|$, and $M_\mu f \in L^2$ because μ satisfies the doubling condition. A similar argument yields (5.23).   Q.E.D.

REMARKS. 1) The L^2 estimates for $e^{t\mathcal{L}}$ proved in this section can be reformulated in terms of the spectral resolution $\{E_\lambda\}$ of the operator $\sqrt{-\mathcal{L}}$, as in [S4]. We omit the details.

2) In [S4] we gave some L^p analogues of Theorems 5.3 and 5.5. It is not clear if all these results can be extended to our context.

We consider next the case when the measure μ is a regular measure supported on a submanifold, and estimates (5.18), (5.19), (5.22), (5.23) can be replaced by exact limits. The situation is more complicated than in the case of Euclidean space. For simplicity of exposition we consider first the case of a curve $\gamma(t)$ and the measure μ given by

$$(5.25) \qquad \int f d\mu = \int f(\gamma(t)) dt.$$

We do not assume that γ is parametrized by arc–length; as we shall see, the arc–length measure is in no sense natural for this problem. We could have included a weighting factor in (5.25), but since we are interested in $e^{t\mathcal{L}}(f d\mu)$ we can incorporate the weighting factor into f.

Now let $\{X_{ij}\}$ be a basis for the Lie algebra of G such that $\delta_t X_{ij} = t^i X_{ij}$. Let $\mathfrak{g}_i = \text{span}\{X_{ij}\}_j$ and $\mathfrak{h}_k = \text{span}\{X_{ij}\}_{j, i \le k}$ and let P_i denote the projection onto $\mathfrak{g}_i$. Note that $\mathfrak{g}_i$ and $\mathfrak{h}_k$ are subspaces, not subalgebras of $\mathfrak{g}$. We will consider the elements $X \in \mathfrak{g}$ as left–invariant vector fields, and write $X(x)$ for the tangent vector in the tangent space at x that the vector field passes through. Thus

$$\mathfrak{h}_1(x) \subseteq \mathfrak{h}_2(x) \subseteq \cdots \subseteq \mathfrak{h}_r(x)$$

is a nested sequence of subspaces of the tangent space at x. Similarly we write $P_i(x)$ for the projection onto $\mathfrak{g}_i(x)$ in the tangent space at x.

We consider Lipschitz curves, so $\gamma'(t)$ is defined a.e. We define the <u>rank</u> of γ to be the minimum k such that $\gamma'(t) \in \mathfrak{h}_k(\gamma(t))$ a.e. We say the curve has <u>constant rank</u> if $P_k(\gamma(t))\gamma'(t) \ne 0$ a.e. (for k equal the rank). The next lemma says that a curve of rank k maintains close contact to the manifold $\gamma(t_0) \exp \mathfrak{h}_k$ for t near t_0, closer than would be the case if we only knew $\gamma'(t_0) \in \mathfrak{h}_k(\gamma(t_0))$ at that point. It may seem at first surprising that we can obtain higher order contact without assuming more differentiability.

LEMMA 5.6. *Let γ be a Lipschitz curve of rank k. Then for a.e. point t_0, we have*

$$(5.26) \qquad \gamma(t_0 + t) = \gamma(t_0) \exp Z(t)$$

where

$$(5.27) \qquad P_i Z(t) = \begin{cases} O(t) & i \le k \\ o(t^{i/k}) & i > k \end{cases}$$

as $t \to 0$.

PROOF. This follows from the generalized Campbell-Baker-Hausdorff-Dynkin (GCBHD) formula [S7]. Since we are working in the case of a nilpotent Lie group, this is a finite sum. Since $\gamma'(t) \in \mathfrak{h}_k(t)$ a.e. we can write

(5.28) $$\gamma'(t) = \sum_{i \leq k} b_{ij}(t) X_{ij}(\gamma(t))$$

with $b_{ij} \in L^\infty$, or equivalently

(5.28') $$\gamma'(t_0 + t) = b(t)(\gamma(t))$$

where $b(t)$ is an L^∞ function taking values in $\mathfrak{h}_k$. Then (GCBHD) says that (5.26) holds with

(5.29) $$Z(t) = \sum_{j=1}^{r} \sum_{\sigma} c(r, \sigma) \int_{T_j(t)} [\cdots [b(s_{\sigma(1)}) b(s_{\sigma(2)})] \cdots] b(s_{\sigma_j})] ds$$

where σ runs over all permutations on j letters, $c(r, \sigma)$ are certain combinatorial coefficients whose exact values are not important here, and $T_j(t)$ denotes the simplex in $\mathbb{R}^j$ given by $0 < s_1 < s_2 \cdots < s_j < t$. Since this is a finite sum it suffices to estimate each term, and we may assume σ is the identity for simplicity of notation. Observe that $[\cdots [b(s_1) b(s_2)] \cdots] b(s_j)] \in \mathfrak{h}_{kj}$ so to obtain (5.27) it is more than sufficient to show

(5.30) $$\int_{T_j} [\cdots [b(s_1) b(s_2)] \cdots] b(s_j)] ds = \begin{cases} O(t) & j = 1 \\ o(t^j) & j > 1, \end{cases}$$

(we in fact obtain $o(t^m)$ in (5.27) where m is the smallest integer greater than or equal to i/k when $i > k$).

Now a straightforward estimate of the integral (5.30) using the boundedness of $b(s)$ and the volume of the simplex gives only $O(t^j)$. To obtain the improvement (for $j \geq 2$) we observe that

$$[b(s_1), b(s_2)] = [b(s_1) - b(s_2), b(s_2)]$$

and hence the integral is bounded by

$$\text{const} \int_{T_j} |b(x_1) - b(s_2)| ds$$

and this is $o(t^j)$ if 0 is a Lebesgue point of $b(s)$. But since b is bounded, almost every point is a Lebesgue point, and so (5.27) holds for a.e. t_0. Q.E.D.

We define a weight function

(5.31) $$w(X) = \int_{-\infty}^{\infty} h_1(\exp tX) dt$$

on $\mathfrak{g}$.

THEOREM 5.7. *Let γ be a finite length Lipschitz curve of constant rank k, and assume $f \in L^p(d\mu)$ for some p, $1 \leq p \leq \infty$ and μ given by (5.25). Then*

(5.32) $$\lim_{t \to 0} t^{(Q-k)/2} e^{t\mathcal{L}}(fd\mu)(\gamma(s)) = f(\gamma(s))w(P_k\gamma'(s))$$

for a.e. s.

REMARK. It is easy to see that

$$\lim_{t \to 0} t^{(Q-k)/2} e^{t\mathcal{L}}(fd\mu)(x) = 0$$

for every x outside the closure of the image of γ.

PROOF. We will show that (5.32) holds for every value s for which Lemma 5.6 holds, $P_k(\gamma'(s)) \neq 0$ and s is a Lebesgue point for $f(\gamma(s))$. Using (5.2), (5.3) and (5.26) we have

$$t^{(Q-k)/2} e^{t\mathcal{L}}(fd\mu)(\gamma(s_0)) = t^{-k/2} \int h_1(\delta_{t^{-1/2}}\gamma(s)^{-1}\gamma(s_0))f(\gamma(s))ds$$

$$= \int h_1(\delta_{t^{-1/2}} \exp(-Z(st^{k/2})))f(\gamma(st^{k/2} + s_0))ds.$$

Now we can write

$$Z(st^{k/2}) = \sum_{i=1}^{r} P_i Z(st^{k/2})$$

and then

$$\delta_{t^{-1/2}} \exp(-Z(st^{k/2})) = \exp\left(-\sum_{i=1}^{r} t^{-i/2} P_i Z(st^{k/2})\right).$$

Now by (5.27) we have

$$\lim_{t \to 0} t^{-i/2} P_i Z(st^{k/2}) = 0 \text{ for } i \neq k$$

while it is easy to see that

$$\lim_{t \to 0} t^{-k/2} P_k Z(st^{k/2}) = sP_k\gamma'(s_0).$$

Thus

$$\lim_{t \to 0} t^{(Q-k)/2} e^{t\mathcal{L}}(fd\mu)(\gamma(s_0))$$

$$= \lim_{t \to 0} f(\gamma(s_0)) \int h_1\left(\exp\left(-\sum_{i=1}^{r} t^{-i/2} P_i Z(st^{k/2})\right)\right)ds$$

$$+ \lim_{t \to 0} \int h_1\left(\exp\left(-\sum_{i=1}^{r} t^{-i/2} P_i Z(st^{k/2})\right)\right)(f(\gamma(st^{k/2} + s_0)) - f(\gamma(s)))ds$$

and the first limit is equal to $f(\gamma(s_0))w(P_k\gamma'(s_0))$ by the dominated convergence theorem. Finally the second limit is equal to zero if s_0 is a Lebesgue point of $f(\gamma(s))$ by standard approximate identity estimates. Q.E.D.

THEOREM 5.8. *Let the hypotheses of Theorem 5.7 hold with $p = 2$, and in addition assume $w(P_k\gamma'(s)) \in L^\infty$. Then*

(5.33)
$$\lim_{t \to 0} t^{(Q-k)/2}\|e^{t\mathcal{L}}(f d\mu)\|_2^2 = 2^{(k-Q)/2} \int |f(\gamma(s))|^2 w(P_k\gamma'(s))ds.$$

PROOF. The proof is essentially the same as the proof of Theorem 5.5. We use (5.24) with $\alpha = k$ and apply the dominated convergence theorem, substituting (5.32) to obtain the limit of the integrand, and using the one–dimensional maximal function of $f(\gamma(s))$ to dominate the integrand.   Q.E.D.

Now we consider the case of a general m–dimensional submanifold, which we assume to be of class C^1. For simplicity we deal with a single coordinate chart, which we denote by $\gamma(s) = \gamma(s_1, \ldots, s_m)$ for s in some bounded open set Ω in $\mathbb{R}^m$, and the measure μ given by

(5.34)
$$\int f d\mu = \int_\Omega f(\gamma(s))ds.$$

We need to define the <u>rank</u> of γ at s, but now this will be a more complicated invariant, consisting of a sequence of integers k_j satisfying

(5.35)
$$r \geq k_1 > k_2 > \cdots > k_\ell \geq 1$$

and dimensions $m_1, \ldots, m_\ell$ satisfying

(5.36)
$$m_1 + \cdots + m_\ell = m$$

defined recursively as follows: first $k_1 = \max\{k : P_k D\gamma \neq 0\}$, $m_1 = \dim\{P_{k_1} D\gamma\}$, where $D\gamma$ denotes the tangent space to γ at s, and $U_1 = \ker P_{k_1}|_{D\gamma}$; then $k_j = \max\{k : P_k U_{j-1} \neq 0\}$, $m_j = \dim\{P_{k_j} U_{j-1}\}$ and $U_j = \ker P_{k_j}|_{U_{j-1}}$.

We say that γ has <u>constant rank</u> if the numbers k_j and m_j are the same for all $s \in \Omega$. In that case, by the implicit function theorem, there exists a basis $v_{ij}(s)$ of $\mathbb{R}^m$, $1 \leq i \leq \ell$, $1 \leq j \leq m_\ell$, such that $\{P_{k_i}v_{ij}(s) \cdot D\gamma(s)\}_{j=1,\ldots,m_i}$ is a basis for $P_{k_i}U_{k_{i-1}}$ (we write $U_0 = D\gamma$). The effective dimension of γ will then be

(5.37)
$$\alpha = \sum_{j=1}^\ell k_j m_j$$

which satisfies $m \leq \alpha \leq Q$.

Now we can define the weighting factor analogous to $w(P_k\gamma'(s))$ that appears in (5.32) and (5.33). We write $|\det v(s)|$ for the volume in $\mathbb{R}^m$ determined by the vectors $v_{ij}(s)$. Then set

(5.38)
$$w(s) = |\det v(s)| \int_{\mathbb{R}^m} h_1\Big(\exp \sum_{i,j} \lambda_{ij} P_{k_i}v_{ij}(s) \cdot D\gamma(s)\Big)d\lambda.$$

Because of the factor $|\det v(s)|$ this is independent of the choice of basis $v_{ij}(s)$.

THEOREM 5.9. *Let γ be a bounded C^1 submanifold of constant rank, and assume $f \in L^p(d\mu)$ for some p, $1 \leq p \leq \infty$ and μ given by (5.34). Then*

(5.39) $$\lim_{t \to 0} t^{(Q-\alpha)/2} e^{t\mathcal{L}} (f d\mu)(\gamma(s)) = f(\gamma(s)) w(s)$$

a.e., where $w(s)$ is given by (5.38).

PROOF. By making a change of variable in a neighborhood of a fixed point $\tilde{s}$ we may arrange to have $v_{ij}(\tilde{s})$ equal to the coordinate vectors in $\mathbb{R}^m$. For simplicity of notation we will present the argument for the case $m = 2$, $\ell = 2$, $j_1 = j_2 = 1$. Thus we have two integers $k_1 > k_2$ such that

(5.40) $$\begin{cases} P_k \frac{\partial \gamma}{\partial s_1}(\tilde{s}) = 0 \text{ if } k > k_1 \\ P_{k_1} \frac{\partial \gamma}{\partial s_1}(\tilde{s}) \neq 0 \end{cases}$$

(5.41) $$\begin{cases} P_k \frac{\partial \gamma}{\partial s_2}(\tilde{s}) = 0 \text{ if } k > k_2 \\ P_{k_1} \frac{\partial \gamma}{\partial s_2}(\tilde{s}) \neq 0. \end{cases}$$

Using the constant rank hypothesis and solving an o.d.e. we can find a curve λ in $\mathbb{R}^m$ passing through $\tilde{s}$ such that $P_k(\gamma \circ \lambda)' = 0$ if $k > k_2$ and $P_{k_2}(\gamma \circ \lambda)' \neq 0$ along the whole curve. By changing variables we can assume the curve is the s_2–axis. Thus we have a strengthened version of (5.41),

(5.41′) $$\begin{cases} P_k \frac{\partial \gamma}{\partial s_2}(\tilde{s} + (0, s_2)) = 0 \text{ if } k > k_2 \\ P_{k_2} \frac{\partial \gamma}{\partial s_2}(\tilde{s} + (0, s_2)) \neq 0 \end{cases}$$

Imitating the argument in the proof of Theorem 5.7 we find

$$t^{(Q-\alpha)/2} e^{t\mathcal{L}} (f d\mu)(\gamma(\tilde{s})) = \iint h_1(\delta_{t^{-1/2}} \exp(-Z(s_1 t^{k_1/2}, s_2 t^{k_2/2}))$$
$$f(\gamma(s_1 t^{k_1/2} + \tilde{s}_1, s_2 t^{k_2/2} + \tilde{s}_2)) ds_1 ds_2$$

where now

$$\gamma(s + \tilde{s}) = \gamma(\tilde{s}) Z(s).$$

We take a path from $\tilde{s}$ to $\tilde{s} + s$ that first moves parallel to the s_2–axis (where we can use (5.41′)) and then parallel to the s_1–axis (we use the analogue of (5.40) which holds throughout a neighborhood of $\tilde{s}$) and adapt the proof of Lemma 5.6 to obtain

$$\lim_{t \to 0} \delta_{t^{-1/2}} \exp(-Z(s_1 t^{k_1/2}, s_2 t^{k_2/2}))$$
$$= \lim_{t \to 0} \exp\left(-\sum_i t^{-i/2} P_i Z(s_1 t^{k_1/2}, s_2 t^{k_2/2})\right)$$
$$= s_1 P_{k_1} \frac{\partial \gamma}{\partial s_1}(\tilde{s}) + s_2 P_{k_2} \frac{\partial \gamma}{\partial s_2}(\tilde{s}).$$

Then (5.39) follows by standard approximate identity arguments for every Lebesgue point of $f(\gamma(s))$.   QED.

THEOREM 5.10. *Let the hypotheses of Theorem 5.9 hold with $p = 2$. Then*

$$(5.42) \qquad \lim_{t \to 0} t^{(Q-\alpha)/2} \|e^{t\mathcal{L}}(fd\mu)\|_2^2 = 2^{(\alpha-Q)/2} \int |f(\gamma(s))|^2 w(ds)ds.$$

PROOF. Same as Theorem 5.8.   Q.E.D.

It is a natural problem to try to extend these results to the heat semigroup $e^{t\mathcal{L}}$ associated to a general Hörmander–type subelliptic operator $\mathcal{L}$. There exist a number of works deriving asymptotic expansions (as $t \to 0$) for the associated heat kernel (see [BA] and [Lea], for example). However, these estimates are all of the form

$$h_t(x,y) \sim ct^{-Q/2}e^{-d(x,y)^2/4t} \quad t \to 0$$

where $d(x,y)$ denotes the associated subRiemannian metric, and they are not sharp enough in the case we have considered to obtai n our results. In particular, they would lead to an incorrect computation of the weighting factors in (5.32) and (5.39).

§6. Self–similar Measures

In this section we will use only the dilation structure of the stratified nilpotent Lie group, and not the lattice subgroup. As in the previous section we assume we are given a sub–Laplacian $\mathcal{L}$. The basis $X_1, \dots, X_{n_1}$ in (5.1) defines an inner product on $\mathbb{R}^{n_1}$, the first component $\mathfrak{g}_1$ of $\mathfrak{g}$, by the requirement that it be an orthonormal basis, and in fact $\mathcal{L}$ depends only on this inner product, and not the particular orthonormal basis chosen. In terms of this inner product we can define the analogue of the orthogonal group on Euclidean space. Let $O(G)$ denote the subgroup of the automorphisms of G which act orthogonally on $\mathfrak{g}_1$. Since each automorphism is determined by its action on $\mathfrak{g}_1$, we can identify $O(G)$ with a subgroup of $O(\mathfrak{g}_1)$, but not every orthogonal transformation on $\mathfrak{g}_1$ extends to an automorphism of G. A priori, the only elements we know are in $O(G)$ are the identity and the map $(x_1, x_2, x_3, \dots, x_r) \to (-x_1, x_2, -x_3, \dots, (-1)^r x_r)$. It is possible to choose the metric on G so that $O(G)$ acts isometrically. Also note that the dilations δ_t commute with $O(G)$.

The <u>similarity group</u> of G is generated by left translations, dilations and $O(G)$. Each similarity S can be written

$$(6.1) \qquad\qquad\qquad Sx = b\delta_\rho Rx$$

where $b \in G$, $\rho > 0$ and $R \in O(G)$. S is <u>contractive</u> if $\rho < 1$. A probability measure μ on G is called <u>self–similar</u> if there exist contractive similarities $S_1, \dots, S_m$ and positive weights $a_1, \dots, a_m$ satisfying

$$(6.2) \qquad\qquad\qquad \sum_{j=1}^{m} a_j = 1$$

such that

$$(6.3) \qquad \mu = \sum_{j=1}^{m} a_j \mu \circ S_j^{-1}.$$

A general result of Hutchinson [H] asserts the existence and uniqueness of solutions to (6.3). The similarities are said to satisfy the <u>open set condition</u> if there exists an open set U such that $S_1 U, \ldots, S_m U$ are disjoint subsets of U. We say the <u>strong open set condition</u> holds if the closures of $S_j U$ are disjoint. A slightly weaker condition is that $\mu(\partial U) = 0$ (since this depends on μ it depends on the weights as well as the similarities). See [LW] for a discussion of these conditions in Euclidean space — all their results extend to our context with essentially the same proof.

Self–similar measures provide examples of the kind of fractal measures discussed in the last section. Given similarities $S_1, \ldots, S_m$ with contraction ratios $\rho_1, \ldots, \rho_m$, there is a set of weights called <u>natural weights</u> which satisfy

$$(6.4) \qquad a_j = \rho_j^{\alpha}, \quad j = 1, \ldots, m$$

for some α (there is a unique α for which (6.2) and (6.4) both hold). If the open set condition holds and the natural weights are used then $\mu = c\mu_\alpha|_E$ for a certain compact set E, and μ is both locally uniformly α–dimensional and quasi–regular. This is proved in [H] and [S1] for Euclidean space, but the same proofs apply. Thus, by the results of the previous section, we know that the expressions

$$(6.5) \qquad H(t) = t^{(Q-\alpha)/2} \|e^{t\mathcal{L}}\mu\|_2^2$$

$$(6.6) \qquad h(t) = t^{-Q-\alpha} \|\mu(B_t(x))\|_2^2$$

are bounded above and below as $t \to 0$.

However, we can assert a lot more, and without assuming that the weights are natural. We choose α to satisfy the identity

$$(6.7) \qquad \sum (a_j)^2 \rho_j^{-\alpha} = 1.$$

Now, a straightforward computation shows that if S is of the form (6.1) then

$$(6.8) \qquad e^{t\mathcal{L}}(\mu \circ S^{-1})(x) = \rho^{-Q} e^{\rho^{-2}t\mathcal{L}}\mu(S^{-1}x)$$

(this uses the fact that the heat kernel is invariant under the action of $0(G)$). Therefore, the self–similarity identity (6.3) tells us that

$$(6.9) \qquad e^{t\mathcal{L}}\mu(x) = \sum_{j=1}^{m} a_j \rho_j^{-Q} e^{\rho_j^{-2}t\mathcal{L}}\mu(S_j^{-1}x).$$

We substitute (6.9) into the definition (6.5) of $H(t)$ and treat the diagonal terms as the main terms and the cross terms as error terms. This yields

$$(6.10) \qquad H(t) = \sum_{j=1}^{m} |a_j|^2 \rho_j^{-1} H(\rho_j^{-2}t) + \sum_{j \neq k} a_j \bar{a}_k E_{jk}(t)$$

where

$$(6.11) \qquad E_{jk}(t) = t^{(Q-\alpha)/2} \langle e^{t\mathcal{L}}(\mu \circ S_j^{-1}), e^{t\mathcal{L}}(\mu \circ S_k^{-1}) \rangle.$$

A similar computation shows that

$$(6.12) \qquad h(t) = \sum_{j=1}^{m} |a_j|^2 \rho_j^{-1} h(\rho_j^{-2}t) + \sum_{j \neq k} a_j \bar{a}_k e_{jk}(t)$$

where

$$(6.13) \qquad e_{jk}(t) = t^{-Q-\alpha} \int \mu \circ S_j^{-1}(B_t(x)) \mu \circ S_k^{-1}(B_t(x)) dx.$$

LEMMA 6.1. *Assume that the open set condition holds and $\mu(\partial U) = 0$. Then there exists $\varepsilon > 0$ such that*

$$(6.14) \qquad e_{jk}(t) = O(t^\varepsilon) \text{ as } t \to 0$$

and

$$(6.15) \qquad E_{jk}(t) = O(t^{\varepsilon/2}) \text{ as } t \to 0.$$

PROOF. Note that we can also write

$$(6.16) \qquad e_{jk}(t) = t^{-Q-\alpha} \iint \lambda(B_t(x) \cap B_t(y)) d\mu \circ S_j^{-1}(x) d\mu \circ S_k^{-1}(y)$$

where λ denotes the Haar measure on G. If the strong open set condition holds then $e_{jk}(t) = 0$ for small enough t. The proof of (6.14) under the weaker hypothesis $\mu(\partial U) = 0$ is due to Lau and Wang [LW] (they give the proof for the case of Euclidean space, but the same proof is valid in our context). We refer the reader to [LW] for the details.

To prove (6.15) from (6.14) we use (6.16) together with the estimate $\lambda(B_t(x) \cap B_t(y)) \geq ct^Q$ if $d(x,y) \leq t/2$ and (6.14) to obtain

$$(6.17) \qquad \iint \chi(d(x,y) \leq r) d\mu \circ S_j^{-1}(x) d\mu \circ S_k^{-1}(x) \leq cr^{\alpha+\varepsilon}$$

for small r (it holds trivially for large r). Now we know

$$h_t(x) \leq c_N t^{-Q/2} (1 + t^{-1/2} d(x,0))^{-N}$$

for any N, and

$$E_{jk}(t) = t^{(Q-\alpha)/2} \int h_{2t}(y^{-1}x) d\mu \circ S_j^{-1}(x) d\mu \circ S_k^{-1}(y)$$

hence

$$E_{jk}(t) \leq c_N t^{-\alpha/2} \int (1 + t^{-1/2} d(x,y))^{-N} d\mu \circ S_j^{-1}(x) d\mu \circ S_k^{-1}(y)$$

$$= N c_N t^{-(\alpha+1)/2} \int_0^\infty (1 + t^{-1/2} r)^{-N-1}$$

$$\left(\iint \chi(d(x,y) \leq r) d\mu \circ S_j^{-1}(x) d\mu \circ S_k^{-1}(y) \right) dr$$

$$\leq c_N' t^{-(\alpha+1)/2} \int_0^\infty (1 + t^{-1/2} r)^{-N-1} r^{\alpha+\varepsilon} dr$$

by (6.17). If we choose N large enough the integral converges and we obtain
(6.15) by a change of variable.   Q.E.D.

We say that $\{\rho_1, \dots, \rho_m\}$ are _exponentially commensurable_ if there exists ρ such that $\rho_j = \rho^{k_j}$ for k_j a positive integer for all j. Otherwise, we say that $\{\rho_1, \dots, \rho_m\}$ are _exponentially incommensurable_.

THEOREM 6.2. _Under the hypotheses of Lemma 6.1, we have_

(6.18) $$\lim_{t \to 0} (H(t) - \tilde{H}(t)) = 0$$

(6.19) $$\lim_{t \to 0} (h(t) - \tilde{h}(t)) = 0$$

where
   (a) $\tilde{H}(t)$ _and_ $\tilde{h}(t)$ _are positive constants if_ $\{\rho_1, \dots, \rho_m\}$ _are exponentially incommensurable._
   (b) $\tilde{H}(t)$ _and_ $\tilde{h}(t)$ _are positive, continuous multiplicatively periodic functions of period_ ρ,

(6.20) $$\tilde{H}(\rho t) = \tilde{H}(t), \quad \tilde{h}(\rho t) = \tilde{h}(t),$$

_if_ $\{\rho_1, \dots, \rho_m\}$ _are exponentially commensurable._

PROOF. The conclusions follow from (6.10) and (6.12), using (6.7) and the Lemma. This may be obtained by using Feller's renewal theorem [Fe] as in [L] of [LW], or directly as in [S3]. The positivity of the functions $\tilde{H}$ and $\tilde{h}$ requires a separate argument, but again it is the same as in the Euclidean case. Q.E.D.

By using a Tauberian theorem we can obtain the same result for

(6.21) $$H_1(t) = t^{Q-\alpha} \| E_{t^{-1}}(\mu) \|_2^2$$

where E_λ denotes the spectral resolution of the operator $\sqrt{-\mathcal{L}}$. In case (a) this is a standard argument ([W] or [T]) but in case (b) it is a bit tricky, see [LW] for details.

The tiles constructed in §2 and §3 are examples of self–similar measures (the measure being a multiple of Haar measure restricted to the tile). More interesting examples may be obtained by considering the faces of the tiles. Theorem 2.4 can be interpreted in this light, with (2.9) being the set theoretic analogue of (6.3). In these examples we use natural weights (in fact all a_j are equal because the ρ_j are all equal to $1/2$).

For some applications it is important to consider self–similar identities of the form (6.3) but without the restriction that the weights be positive (even condition (6.2) is not always satisfied). In this greater generality we allow distributional solutions, but to keep the problem manageable we require the distribution to have compact support.

In Euclidean space, the problem of existence and uniqueness for such self–similar distributions of compact support was solved in [S3] using Fourier transform methods (many special cases can be handled by other methods, as in [DL]). Surprisingly, the same Fourier transform methods will work for step 2 nilpotent Lie groups using the Euclidean Fourier transform, not the group Fourier transform.

In the notation of §3 our group $G = \mathbb{R}^{n_1} \times \mathbb{R}^{n_2}$ with elements $x = (x_1, x_2)$ and group law

$$(6.22) \qquad (x_1, x_2) \circ (x_1', x_2') = (x_1 + x_1', x_2 + x_2' + F_2(x_1, x_1'))$$

where F_2 is a bilinear map from $\mathbb{R}^{n_1} \times \mathbb{R}^{n_1}$ to $\mathbb{R}^{n_2}$. We will assume for simplicity that G is __stratified__, meaning that the image of F_2 generates $\mathbb{R}^{n_2}$ (if not we can split off a complement of the image of F_2 as a direct summand). The orthogonal group $O(G)$ can be described as follows. Let R_1 be an orthogonal matrix in $O(\mathbb{R}^{n_1})$. If there exists a linear transformation R_2 on $\mathbb{R}^{n_2}$ such that

$$(6.23) \qquad\qquad F_2(R_x x_1, R_1 x_1') = R_2 F_2(x_1, x_1')$$

for all x_1 and x_1' in $\mathbb{R}^{n_1}$ then

$$(6.24) \qquad\qquad R(x_1, x_2) = (R_1 x_1, R_2 x_2)$$

is in $O(G)$. It follows from (6.23) the orbits of R_2^k for $k \in \mathbb{Z}$ are bounded, which by linear algebra implies that we can choose an inner product on $\mathbb{R}^{n_2}$ for which R_2 is orthogonal. We assume, without loss of generality, that this is the standard inner product.

If T is a distribution of compact support on G, we define the Euclidean Fourier transform $\hat{T}(\xi_1, \xi_2)$ by

$$(6.25) \qquad\qquad \hat{T}(\xi_1, \xi_2) = \langle T, e^{i(x_1 \cdot \xi_1 + x_2 \cdot \xi_2)} \rangle.$$

For any similarity S we define

$$(6.26) \qquad\qquad \langle T \circ S^{-1}, \varphi \rangle = \langle T, \varphi \circ S \rangle$$

to be consistent with the definition for measures (note that this is not consistent with the usual definition for functions).

LEMMA 6.3. *Let S be given by (6.1) and define $Q : \mathbb{R}^{n_1} \to \mathbb{R}^{n_2}$ by*

(6.27) $$Q x_1 = F_2(b_1, R_1 x_1).$$

Then

(6.28) $$(T \circ S^{-1})\hat{\,}(\xi_1, \xi_2) = e^{i(b_1 \cdot \xi_1 + b_2 \cdot \xi_2)} \hat{T}(\rho(R_1^* \xi_1 + Q^* \xi_2), \rho^2 R_2^* \xi_2).$$

PROOF. Since

$$S(x_1, x_2) = (b_1 + \rho R_1 x_1, b_2 + \rho^2 R_2 x_2 + \rho F_2(b_1, R_1 x_1))$$

we have

$$(T \circ S^{-1})\hat{\,}(\xi_1, \xi_2) = \left\langle T, e^{i(b_1 + \rho R_1 x_1) \cdot \xi_1 + i(b_2 + \rho^2 R_2 x_2 + \rho F_2(b_1, R_1 x_1)) \cdot \xi_2} \right\rangle$$

and (6.28) follows by rearranging terms.   Q.E.D.

Now the point is that if we abbreviate (6.28) as

(6.28′) $$T \circ S^{-1}(\xi) = e^{ib \cdot \xi} \hat{T}(M\xi)$$

where M has block–upper–triangular form,

(6.29) $$M \begin{pmatrix} \xi_1 \\ \xi_2 \end{pmatrix} = \begin{pmatrix} \rho R_1^* & \rho Q^* \\ 0 & \rho^2 R_2^* \end{pmatrix} \begin{pmatrix} \xi_1 \\ \xi_2 \end{pmatrix}$$

then for $\rho < 1$, M is contractive, its eigenvalues being ρ and ρ^2. Thus

(6.30) $$\|M^k\| \le c\rho^k$$

for some c and all $k \ge 1$.

If we substitute (6.28′) into the self–similarity identity (6.3) we obtain

(6.31) $$\hat{T}(\xi) = \sum_{j=1}^{r} a_j e^{ib_j \cdot \xi} \hat{T}(M_j \xi).$$

To solve (6.3) for a distribution of compact support is equivalent to solving (6.31) for an entire function of exponential type. In fact, the proof of Lemma 2.1 in [S3] shows that any entire analytic solution of (6.31) is automatically of exponential type.

We solve (6.31) on the level of formal power series about the origin. We consider $\xi = (\xi_1, \xi_2)$ as a variable in $\mathbb{R}^{n_1} \times \mathbb{R}^{n_2} = \mathbb{R}^{n_1 + n_2}$ and let $\mathcal{P}_k$ denote the space of polynomials homogeneous of degree k (this is the usual homogeneity with ξ_1 and ξ_2 variables treated equally). The linear transformations M_j acts on $\mathcal{P}_k$ via $p_k(\xi) \to p_k(M_j \xi)$, and we denote this action by $M_j^{(k)}$. We define a sequence of algebraic conditions

$$(A_k) \quad I - \sum_{j=1}^{m} a_j M_j^{(k)} \text{ is not invertible on } \mathcal{P}_k.$$

Note that (A_0) is just (6.2).

THEOREM 6.4. *The space of distributions of compact support satisfying* (6.3) *is at most finite dimensional, and non–trivial solutions exist if and only if one of the conditions* (A_k) *holds. If* (A_k) *holds only for* $k = 0$, *the space of solutions is one–dimensional and there is a unique solution with* $\hat{T}(0) = 1$. *In general, if* k' *is the largest value of* k *for which* (A_k) *holds, the dimension of the space of solutions is at least the nullity of* $I - \sum a_k M_j^{(k)}$ *and at most the dimension of* $\sum_{k \leq k'} \mathcal{P}_k$.

PROOF. Because the block–upper-triangular form (6.29) is the same for all the matrices M_j, we can choose a new norm on $\mathbb{R}^{n_1} \times \mathbb{R}^{n_2}$ of the form $\lambda |\xi_1|^2 + |\xi_2|^2$ for λ large enough such that all the matrices M_j are contractive with respect to this norm, $\|M_j\| \leq 1 - \varepsilon$. We make a change of variable in $\mathbb{R}^{n_1} \times \mathbb{R}^{n_2}$ so that this becomes the standard norm. Then

$$(6.32) \qquad\qquad \|M_j^{(k)}\| \leq (1 - \varepsilon)^k$$

for the usual norm on $\mathcal{P}_k$. It is now clear that (A_k) can only hold for a finite number of k's.

The proof is now the same as the proof of Theorem 2.2 in [S3]. If $\hat{T}$ is analytic then write $\hat{T}(\xi) = \sum_{k=0}^{\infty} p_k(\xi)$ for its convergent power series expansion with $p_k \in \mathcal{P}_k$. Then (6.31) is equivalent to

$$(6.33) \qquad p_k(\xi) - \sum_{j=1}^{m} a_j p_k(M_j \xi) = \sum_{j=1}^{m} \sum_{\ell=1}^{k} a_j \frac{(ib_j \cdot \xi)^\ell}{\ell!} p_{k-\ell}(M_j \xi).$$

The failure of (A_k) is thus equivalent to p_k being uniquely determined by p_ℓ for $\ell < k$. The conclusions of the theorem are then straightforward on the level of formal power series solutions to (6.31). But the form of (6.33) and the estimates (6.32) easily imply that the radius of convergence is infinite.   Q.E.D.

§7. Multiperiodic Functions

Let G be as in §3. We are interested in functions of the form

$$(7.1) \qquad\qquad F(x) = \prod_{k=1}^{\infty} f(\delta_{\rho^k} x)$$

for some $\rho < 1$ where f is periodic under Γ and

$$(7.2) \qquad\qquad f(0) = 1.$$

Note that these functions satisfy the identity

$$(7.3) \qquad\qquad F(x) = f(\delta_\rho x) F(\delta_\rho x).$$

We call such functions <u>multiperiodic</u> because they are built out of periodic functions using the multiplicative structure of the dilations. We do not intend to make this the definition of the term "multiperiodic", because it should be used for a larger class of functions. Included under this broader definition should be functions satisfying

(7.4)
$$F(x) = \sum_{j=1}^{m} f_j(\delta_{\rho_j} x) F(\delta_{\rho_j} x)$$

where the f_j's are periodic (perhaps with respect to different lattice subgroups) and all $\rho_j < 1$. More generally, one should allow almost periodic functions in the sense of Auslander and Brezin [AB], and actions of the orthogonal group $O(G)$. The motivation for studying these functions is that in the Euclidean case they include Fourier transforms of certain self–similar measures and distributions ([JRS]). In the context of nilpotent Lie groups we do not have such an interpretation.

Our goal is to generalize the results of [JRS] on the asymptotic behavior of multiperiodic functions. To this end we consider the sequence

$$\frac{1}{n} \log |F(\delta_{\rho^{-n}} x)|$$

and study the behavior as $n \to \infty$ (for $x \neq 0$, the sequence $\delta_{\rho^{-n}} x$ tends to infinity in G). It follows easily from the definition that

(7.5)
$$\frac{1}{n} \log |F(\delta_{\rho^{-n}} x)| = h_n(x) + \frac{1}{n} \log |F(x)|$$

where

(7.6)
$$h_n(x) = \frac{1}{n} \sum_{j=0}^{n-1} \log |f(\delta_{\rho^{-j}} x)|$$

so the asymptotic behavior of $h_n(x)$ is the same as $\frac{1}{n} \log |F(\delta_{\rho^{-n}} x)|$ at all points where $F(x)$ is finite and non–zero.

Since f is assumed to be Γ periodic, we can regard it as a function on any fundamental domain for Γ. This simplest choice is to take a tile T given by (3.7) with all $A_{ij}(x) = 0$. Note that the Haar measure of T is $\Pi_{\ell=1}^{r}(k_\ell)^{-n_\ell}$. By Lemma 3.1 we can define a projection $P : G \to T$ that sends x to the unique point γx in T for some $\gamma \in \Gamma$. In fact γ is given inductively by (3.8) and Px is then simply

(7.7)
$$(Px)_{\ell j} = k_\ell^{-1} \langle k_\ell(x_{\ell j} + F_{\ell j}(\gamma, \lambda)) \rangle$$

where $\langle \ \rangle$ denotes the fractional part. Then we have

(7.8)
$$h_n(x) = \frac{1}{n} \sum_{j=0}^{n-1} \log |f(P\delta_{\rho^{-j}} x)|$$

and the right side resembles a Riemannian sum for the integral of $\log |f|$ over T. To make this precise we need show that the sequence of points $P\delta_{\rho^{-j}} x$ is uniformly distributed on T. Recall that a sequence $\{\lambda_j\}$ of points is said to be uniformly distributed on T if

(7.9)
$$\lim_{n\to\infty} \frac{1}{n} \sum_{j=0}^{n-1} f(\lambda_j) = \prod_{\ell=1}^{r} (k_\ell)^{n_\ell} \int_T f(x)dx$$

for every Riemann integrable function f on T. By a theorem of Weyl it suffices to verify

(7.10)
$$\lim_{n\to\infty} \frac{1}{n} \sum_{j=0}^{n-1} e^{2\pi i \sum_{\ell,q} k_\ell m_{\ell q}(\lambda_j)_{\ell q}} = 0$$

for all integral $m_{\ell q}$ not identically zero.

LEMMA 7.1. *Fix $\rho < 1$. For almost every x, the sequence $P\delta_{\rho-j}x$ is uniformly distributed on T.*

PROOF. We generalize a method of proof due to Koksma [Ko]. We first obtain an L^2 version of (7.10). Let

(7.11)
$$s_n(x) = \frac{1}{n} \sum_{j=0}^{n-1} e^{2\pi i \sum_{\ell,q} k_\ell m_{\ell q}(P\delta_{\rho-j}x)_{\ell q}}.$$

Then

$$\|s_n\|_2^2 = \int_T |s_n(x)|^2 dx = \frac{1}{n}\Pi(k_\ell)^{-n_\ell}$$

(7.12)
$$+ \frac{1}{n^2} \sum_{0 \le j \ne k < n} \int_T e^{2\pi i \sum_{\ell,q} k_\ell m_{\ell q}[(P\delta_{\rho-j}x)_{\ell q} - (P\delta_{\rho-k}x)_{\ell q}]} dx$$

We need to estimate the integrals in (7.12). Let $\ell_0 q_0$ be chosen so that $m_{\ell_0 q_0} \ne 0$ and ℓ_0 is as large as possible. Then

$$e^{2\pi i k_{\ell_0} m_{\ell_0 q_0}[(P\delta_{\rho-j}x)_{\ell_0 q_0} - (P\delta_{\rho-k}x)_{\ell_0 q_0}]}$$

$$= e^{2\pi i m_{\ell_0 q_0}(\langle k_{\ell_0}(\rho^{-j\ell_0}x_{\ell_0 q_0} + F_{\ell_0 q_0}(\gamma, \delta_{\rho-j}x))\rangle - \langle k_{\ell_0}(\rho^{-k\ell_0}x_{\ell_0 q_0} + F_{\ell_0 q_0}(\gamma, \delta_{\rho-k}x))\rangle)}$$

$$= e^{2\pi i m_{\ell_0 q_0} k_{\ell_0}(\rho^{-j\ell_0} - \rho^{-k\ell_0})x_{\ell_0 q_0}}$$

$$\cdot e^{2\pi i m_{\ell_0 q_0} k_{\ell_0}(F_{\ell_0 q_0}(\gamma, \delta_{\rho-j}x) - F_{\ell_0 q_0}(\gamma, \delta_{\rho-k}x))}$$

and none of the other factors involves the variable $x_{\ell_0 q_0}$. When we do the integration with respect to this variable we produce a factor of $|\rho^{-j\ell_0} - \rho^{-k\ell_0}|^{-1}$. All the other factors have absolute value one so (7.12) yields the estimate

(7.13)
$$\|s_n\|_2^2 \le \frac{c}{n} + \frac{c}{n^2} \sum_{0 \le j < k < n} |\rho^{-j\ell_0} - \rho^{-k\ell_0}|^{-1} = O(\frac{1}{n}).$$

This implies that $s_{n^2}(x) \to 0$ a.e. But from the form of (7.11) we have that if $n^2 \le N \le (n+1)^2$ then

$$|s_n(x) - s_N(x)| = \left|\left(\frac{1}{N} - \frac{1}{n^2}\right)\sum_1^{n^2} w_j + \frac{1}{N}\sum_{n^2+1}^N w_j\right|$$

with $|w_j| = 1$ so

$$|s_n(x) - s_N(x)| \le \frac{c}{N}.$$

Thus the convergence of the subsequence implies convergence of the sequence.
Q.E.D.

THEOREM 7.2. *Fix $\rho < 1$, or almost every x,*

(7.14) $$\lim_{n\to\infty} h_n(x) = \prod_{\ell=1}^r (k_\ell)^{n_\ell} \int_T \log |f(x)| dx$$

if f is continuous and non–vanishing.

PROOF. Since $\log |f|$ is continuous we apply the Lemma to (7.8).   Q.E.D.

In the special case the ρ^{-1} is an integer we can say more. Then the mapping $P\delta_{\rho^{-1}} : T \to T$ is measure preserving and ergodic. Indeed, $\delta_{\rho^{-1}}$ has Jacobian ρ^{-Q} and each point in T has exactly ρ^{-Q} pre–images under $P\delta_{\rho^{-1}}$, so $\lambda(P\delta_{\rho^{-1}}(E)) = \lambda(E)$. The proof of ergodicity follows by standard arguments since $P\delta_{\rho^{-1}}$ is locally expanding. Thus the individual ergodic theorem yields (7.14) under the weak hypothesis $\log |f| \in L^p(T)$, $1 \le p \le \infty$.

As pointed out in [JRS], the almost everywhere convergence in Theorem 7.2 is not everywhere convergence, and the functions $h_n(x)$ exhibit a very complicated behavior. For ρ^{-1} an integer, they have a kind of asymptotic self–similarity with respect to the transformation $P\delta_{\rho^{-1}}$, or any of its powers.

THEOREM 7.3. *Let ρ^{-1} be an integer. If f is continuous and non–vanishing, then*

(7.15) $$h_n(P\delta_{\rho^{-1}}x) - h_n(x) \to 0 \text{ uniformly as } n \to \infty,$$

in fact at a rate $O(1/n)$. If $\log |f|$ is in $L^p(T)$, then the convergence is in $L^p(T)$ norm.

PROOF. Since f is periodic

$$h_n(P\delta_{\rho^{-1}}x) = \frac{1}{n}\sum_{j=1}^n \log |f(P\delta_{\rho^{-j}}x)|$$

hence

$$h_n(P\delta_{\rho^{-1}}x) - h_n(x) = \frac{1}{n}(\log |f(P\delta_{\rho^{-j}}x)| - \log |f(x)|)$$

and the result follows easily.   Q.E.D.

References

[AB]   L. Auslander and J. Brezin, *Uniform distributions in solvmanifolds*, Advances in Math **7** (1971), 111–144.

[B]   C. Bandt, *Self–similar sets 5, integer matrices and fractal tilings of* $\mathbb{R}^n$, Proc. Amer. Math. Soc. **112** (1991), 549–562.

[BA]   G. Ben Arous, *Développement asymptotique du noyan de la chaleur hypoélliptique hors du cut–locus*, Ann. Scient. Ec. Norm. Sup. **21** (1988), 307–331.

[DL]   I. Daubechies and J. C. Lagarias, *Two–scale difference equations I. Existence and global regularity of solutions*, SIAM J. Math. Anal. **22** (1991), 1388–1410.

[E]   H. G. Eggleston, *Convexity*, Cambridge University Press, 1958.

[F1]   K. J. Falconer, *The geometry of fractal sets*, Cambridge University Press, 1985.

[F2]   K.J. Falconer, *Fractal geometry: mathematical foundations and applications*, John Wiley, 1990.

[Fe]   W. Feller, *An introduction to probability theory and its applications*, vol. 2, 2nd edition, John Wiley, 1971.

[FS]   G. B. Folland and E. M. Stein, *Hardy spaces on homogeneous groups*, Princeton University Press, Math. Notes **28** (1982).

[G]   B. Gaveau, *Principe de moindre action, propagation de la chaleur et estimées sous élliptiques sur certains groupes nilpotents*, Acta Math. **139** (1977), 95–153.

[GH]   K. Gröchenig and A. Haas, *Self–similar lattice tilings*, preprint.

[H]   J. E. Hutchinson, *Fractals and self–similarity*, Indiana University Math. J. **30** (1981), 713–747.

[JRS]   P. Janardhan, D. Rosenblum and R. S. Strichartz, *Numerical experiments in Fourier asymptotics of Cantor measures and wavelets*, preprint.

[K]   R. Kenyon, *Self–replicating tilings*, preprint.

[Ko]   J. F. Koksma, *Ein mengentheoretischer Satz über die Gleichverteilung modulo Eins*, Compositio Math. **2** (1935), 250–258.

[KR]   A. Koranyi and H. M. Reimann, *Foundations for the theory of quasiconformal mappings on the Heisenberg group*, preprint.

[KV]   A. Koranyi and S. Vagi, *Singular integrals in homogeneous spaces and some problems of classical analysis*, Ann. Scuola Norm. Sup. Pisa **25** (1971), 575–648.

[L]   K.-S. Lau, *Fractal measures and mean p–variations*, J. Functional Anal. (to appear).

[LW]   K.-S. Lau and J. Wang, *Mean quadratic variations and Fourier asymptotics of self–similar measures*, preprint.

[Lea]   R. Léandre, *Développement asymptotique de la densité d'une diffusion dégenerée*, Forum Math. **4** (1992), 45–75.

[Le]   P. G. Lemarié, *Bases d'ondelettes sur les groupes de Lie stratifiés*, Bull. Soc. Math. France **117** (1989), 211–232.

[M]   B. B. Mandelbrot, *The fractal geometry of nature*, W. H. Freeman & Co., 1982.

[Mi]   J. Mitchell, *On Carnot–Caratheodory metrics*, J. Diff. Geom. **21** (1985), 35–45.

[NSW]   A. Nagel, E. M. Stein and S. Wainger, *Balls and metrics defined by vector fields I*, Acta Math. **155** (1985), 103–147.

[SW]   E. Sawyer and R. L. Wheeden, *Weighted inequalities for fractional integrals on Euclidean and homogeneous spaces*, Amer. J. Math. (to appear).

[S1]   R. S. Strichartz, *Fourier asymptotics of fractal measures*, J. Functional Anal. **89** (1990), 154–187.

[S2]   _____, *Self–similar measures and their Fourier transforms I*, Indiana University Math. J. **39** (1990), 797–817.

[S3]   _____, *Self–similar measures and their Fourier transforms II,*, Trans. Amer. Math. Soc. (to appear).

[S4]   _____, *Spectral asymptotics of fractal measures on Riemannian manifolds*, J. Functional Anal. **102** (1991), 176–205.

[S5]   _____, *Wavelets and self–affine tilings* (to appear).

[S6]   _____, *Sub–Riemannian geometry*, J. Diff. Geom. **24** (1986), 221–263, Corrections **30** (1989) 595–596.

[S7]   _____, *The Campbell–Baker–Hausdorff–Dynkin formula and solutions of differential equations*, J. Functional Anal. **72** (1987), 320–345.

[T]   M. Taylor, *Pseudodifferential operators*, Princeton University Press, 1981.

[Th]   W. P. Thurston, *Groups, tilings, and finite state automata*, Lectures notes, Summer meeting of the Amer. Math. Soc., Boulder, 1989.

[W]   N. Wiener, *The Fourier integral and certain of its applications*, Dover, New York, 1933.

MATHEMATICS DEPARTMENT, WHITE HALL, CORNELL UNIVERSITY, ITHACA, N.Y. 14853

Contemporary Mathematics
Volume **140**, 1992

A Kinematic Formula and Analogues of Hadwiger's Theorem in Space

JIAZU ZHOU

ABSTRACT. We first estimate the integral of the total square curvature with respect to kinematic density dg for $\mathbb{R}^3$ *(Theorem 1)* by directly using the methods of differential geometry. Then we derive a kinematic formula and use it to deduce a sufficient condition for one domain to contain another in $\mathbb{R}^3$ *(Theorem 2)*. A sufficient condition for one convex domain to contain another in [**6**] is an easy consequence *(Theorem 3)* of our Theorem 2.

1. Introduction

Let M, N be two submanifolds in a homogeneous space and I an integral invariant of the submanifold $M \cap gN$. Then many works in integral geometry have been concerned with computing integrals of the following type

$$(1) \qquad \int_{\{g \in G: M \cap gM \neq \emptyset\}} I(M \cap gN) \, dg,$$

where dg is the kinematic density. For example in the case that G is the group of rigid motions in n-dimensional Euclidean space $\mathbb{R}^n$, M and N are submanifolds of $\mathbb{R}^n$ and $I(M \cap gN) = Vol(M \cap gN)$ then evaluation of (1) leads to the formulas of Poincaré, Blaschke, Santaló and others (sees [**1**]). R. Howard [**10**] obtained a kinematic formula for $I(M \cap gN) = Vol(M \cap gN)$ in homogeneous space. If $I(M \cap gN) = \chi(M \cap gN)$, where $\chi(\cdot)$ is the Euler-Poincaré characteristic of the intersection $M \cap gN$ of the domains M and N in $\mathbb{R}^n$ with smooth boundary, then (1) leads to S. S. Chern's kinematic fundamental formula [**2**].

Hadwiger (see [**1, 14, 15**]) (1941) obtained some sufficient conditions for one domain to contain another in the 2-dimensional plane $\mathbb{R}^2$ by using kinematic

1991 *Mathematics Subject Classification.* Primary 52A22, 53C05; Secondary 51M16.

Key words and phrases. Integral geometry, kinematic formula, kinematic measure, kinematic density, convex domain, mean curvature, Euler-Poincaré characteristic.

The paper is in final form and no version of it will be submitted for publication elsewhere

formulas. Since then many mathematicians have been interested in getting sufficient conditions to insure that a given domain D_1 of surface area F_1, bounded
by a simple piecewise smooth boundary ∂D_1, of volume V_1 may be moved 'inside' a second domain D_0 of surface area F_0, bounded by a simple piecewise
smooth boundary ∂D_0, of volume V_0 by some kinematic formulas in $\mathbb{R}^n (n \geq 3)$.
The type of condition sought is meaningful if it is 'intrinsic' or 'extrinsic', i.e.,
it just involves volumes V_0, V_1, surface areas F_0, F_1 of boundaries ∂D_0, ∂D_1,
some invariants determined by D_0, D_1, curvatures and the integrals of curva-
tures of ∂D_0, ∂D_1. Ren Delin [3] (1986) obtained other sufficient conditions in
$\mathbb{R}^2$ (which are different from Hadwiger's). It appears to be harder to generalize
Hadwiger's and Ren's results to the situation of $\mathbb{R}^n (n \geq 3)$. Before the works
of [5,6,7,8,12], there was no general result for $\mathbb{R}^n (n \geq 3)$ (not even a restricted
result, with strong topological condition imposed, was available). By restricting
the domains involved to the convex category, the author [6,7,8] and Zhang [5]
obtained some sufficient conditions which are the generalizations of Hadwiger's
theorem. Goodey [9] obtained a related result by putting some strong topological
restrictions to the convex domains involved and their intersection $\partial D_0 \cap g\partial D_1$.

In this paper, we restrict our discussions to $\mathbb{R}^3$ for there the situation is most
concrete and useful. First we consider two compact surfaces M_0 and M_1. Let
M_0 be fixed one, gM_1 be moving one under the rigid motion g, and let dg be
the kinematic density for $\mathbb{R}^3$. The intersection $M_0 \cap gM_1$ is usually a union of
closed curves. By using some known results of classical differential geometry
(e.g., Meusnier's and Euler's theorems) we get the kinematic formula (Theorem
1) for the total square curvature of the curve $M_0 \cap gM_1$ which was also proved by
C-S. Chen [1,4] in 1972 by different means. Then by using the resulting formula
obtained, i.e., the kinematic formula for the total square curvature $\int\limits_{M_0 \cap gM_1} \kappa^2 ds$,
S. S. Chern's kinematic fundamental formula and other known formulas, we get
a sufficient condition (Theorem 2) for one domain to contain another which is
an analogue of Hadwiger's theorem in space $\mathbb{R}^3$. The domains we discuss here
are only assumed to be connected with simple boundaries of class C^2 and not
restricted to either be convex or have other topological properties. One of our
early results (Theorem 3), i.e., a sufficient condition for one convex domain to
contain another [6], is an easy consequence of our results in this paper. In
[13], we are going to generalize the kinematic formula in this paper to higher
dimensional Euclidean space and will derive the general analogues of Hadwiger's
theorem in $\mathbb{R}^n (n \geq 3)$.

We would like to thank Professor Eric Grinberg for his teaching, support and
discussions. We are indebted to Professor E. Lutwak for some discussions during
his visiting to Temple. We also like to thank Professor C. C. Hsiung, Professor
Ren Delin and Professor Yang Wenmao for their support and encouragement.
We appreciate Professor Paul Goodey for his interest and comments. We would
also like to express our sincere thanks to Professor R. Howard for his interest,
comments and helpful suggestions.

2. Main results

Let M be a closed surface of class C^2 in 3-dimensional Euclidean space $\mathbb{R}^3$. Assume K, H are, respectively, Gaussian curvature, mean curvature of M. The total Gaussian curvature, the total mean curvature, the total square mean curvature are, respectively, defined by

$$(2) \qquad \tilde{K} = \int_M K \, dv, \quad \tilde{H} = \int_M H \, dv, \quad \tilde{H}^{(2)} = \int_M H^2 \, dv.$$

THEOREM 1. *Let M_0 and M_1 be two closed surfaces of class C^2 in $\mathbb{R}^3$. Assume that M_0 is fixed, gM_1 is moving under the rigid motion g. Let dg be the kinematic density for $\mathbb{R}^3$, so normalized that the measure of all positions about a point is $8\pi^2$. Let κ_{C_g} denote the curvature of the intersection curve $M_0 \cap gM_1$. Assume that F_i is the surface area, and $\tilde{K}_i$, $\tilde{H}_i$, $\tilde{H}_i^{(2)} (i = 0, 1)$ are the total Gaussian curvature, the total mean curvature and the total square mean curvature of M_i, respectively. Then we have* *

$$(3) \qquad \int_{\{g: M_0 \cap gM_1 \neq \emptyset\}} \left(\int_{M_0 \cap gM_1} \kappa_{C_g}^2 \, ds \right) dg$$

$$= 2\pi^3 (3\tilde{H}_0^{(2)} - \tilde{K}_0) F_1 + 2\pi^3 (3\tilde{H}_1^{(2)} - \tilde{K}_1) F_0,$$

where ds is the arc-element of the intersection curve $C_g = M_0 \cap gM_1$, K_i and H_i are Gaussian curvature and mean curvature of M_i, respectively.

THEOREM 2. *Let D_0 and D_1 be two connected domains in $\mathbb{R}^3$ bounded by the simple surfaces ∂D_0 and ∂D_1, which we assume to be of class C^2. Moreover, we assume that D_0 and D_1 are such that for all $g \in G$, the group of rigid motions in $\mathbb{R}^3$, the Euler-Poincaré characteristic $\chi(D_0 \cap gD_1)$ of the intersection $D_0 \cap gD_1$ is at most N_0, a finite integer. Let V_i $(i = 0, 1)$, F_i, $\chi(D_i)$, $\tilde{K}_i$, $\tilde{H}_i$ and $\tilde{H}_i^{(2)}$ be the volume of D_i, the surface area of D_i, the Euler-Poincaré characteristic of D_i, the total Gaussian curvature of ∂D_i, the total mean curvature and the total square mean curvature of ∂D_i, respectively. Then a sufficient condition for D_0 to enclose D_1 or for D_1 to enclose D_0 is*

$$(4) \quad 8\pi(V_0\chi(D_1) + V_1\chi(D_0)) + 2(F_0\tilde{H}_1 + F_1\tilde{H}_0)$$

$$- N_0\pi\{F_0F_1[(3\tilde{H}_0^{(2)} - \tilde{K}_0)F_1 + (3\tilde{H}_1^{(2)} - \tilde{K}_1)F_0]\}^{\frac{1}{2}} > 0.$$

Moreover,
(i) if $V_0 > V_1$ or $F_0 > F_1$, then D_1 can be enclosed in D_0;
(ii) if $V_1 > V_0$ or $F_1 > F_0$, then D_1 can enclose D_0.

---

*C-S. Chen first proved this formula in 1972 by a different method (see [4]).

Formula (4) comes from estimating the kinematic measure of one domain moving into another under the group G of rigid motions in $\mathbb{R}^3$, i.e.,

(5) $\quad m\{g \in G : gD_1 \subseteq D_0 \quad \text{or} \quad gD_0 \subseteq D_1\}$

$$= \int_{\{g:D_0 \cap gD_1 \neq \emptyset\}} dg - \int_{\{g:\partial D_0 \cap g\partial D_1 \neq \emptyset\}} dg$$

$$\geq \frac{1}{N_0}[8\pi^2(V_0\chi(D_1) + V_1\chi(D_0)) + 2\pi(F_0\tilde{H}_1 + F_1\tilde{H}_0)]$$

$$- \pi^2\{F_0F_1[(3\tilde{H}_0^{(2)} - \tilde{K}_0)F_1 + (3\tilde{H}_1^{(2)} - \tilde{K}_1)F_0]\}^{\frac{1}{2}}.$$

3. Preliminaries and the proof of the theorems

Let M_i $(i = 0, 1)$ be hypersurfaces of class C^2 in $\mathbb{R}^n$, dT_i denote the kinematic densities on M_i, dT_{01} denote the kinematic density on $M_0 \cap gM_1$. Then we have the basic formula (see [1] pp.262 (15.35))

(6) $\qquad\qquad dT_{01} \wedge dg = (\sin^{n-1} \phi)\, d\phi \wedge dT_0 \wedge dT_1,$

where dg is the kinematic density for $\mathbb{R}^n$, ϕ is the angle between M_0 and M_1 at $x \in M_0 \cap gM_1$, i.e., for the frames

$$(x; e_1, \cdots, e_{n-2}, e_{n-1}, e_n), \quad (x; e_1, \cdots, e_{n-2}, e'_{n-1}, e'_n)$$

in $\mathbb{R}^n$.

Now let M_i be the boundaries ∂D_i of domains D_i in $\mathbb{R}^3$. Then the kinematic density on the intersection curve $C_g = \partial D_0 \cap g\partial D_1$ becomes $dT_{01} = ds$, the arc-element, and the kinematic densities on ∂D_i are

(7) $\qquad\qquad dT_i = dv_i \wedge d\phi_i,$

where dv_i is the volume element of ∂D_i at $x \in \partial D_0 \cap g\partial D_1$, ϕ_i is the angle between tangent vector dx and any another fixed tangent vector in $T_x(\partial D_i)$ of ∂D_i. In fact, we can choose the fixed tangent vector as the principal direction in $T_x(\partial D_i)$. Then (6) becomes

(8) $\qquad\qquad ds \wedge dg = (\sin^2 \phi)\, d\phi \wedge dv_0 \wedge d\phi_0 \wedge dv_1 \wedge d\phi_1.$

By using Meusnier's theorem [11], the laws of cosine and sine to the intersection curve at $x \in C_g$ we get the following

LEMMA. *Let M_i $(i = 0, 1)$ be two surfaces of class C^2 in $\mathbb{R}^3$, $\kappa_n^{(i)}$ be the normal curvatures of M_i, κ_{C_g} be the curvature of the intersection curve $C_g = M_0 \cap gM_1$, then we have*

(9) $\qquad\qquad \kappa_{C_g}^2 \sin^2 \phi = \left(\kappa_n^{(0)}\right)^2 + \left(\kappa_n^{(1)}\right)^2 - 2\kappa_n^{(0)}\kappa_n^{(1)} \cos \phi,$

where ϕ is the angle between surfaces M_0 and M_1.

Proof of Theorem 1. By (8) and (9) we have

$$
(10) \quad \int_{\{g:\partial D_0 \cap g\partial D_1 \neq \emptyset\}} \left(\int_{\partial D_0 \cap g\partial D_1} \kappa_{C_g}^2 \, ds \right) dg
$$

$$
= \int [(\kappa_n^{(0)})^2 + (\kappa_n^{(1)})^2 - 2\kappa_n^{(0)}\kappa_n^{(1)} \cos \phi] \, d\phi \wedge dv_0 \wedge d\phi_0 \wedge dv_1 \wedge d\phi_1
$$

$$
= \pi \int ((\kappa_n^{(0)})^2 + (\kappa_n^{(1)})^2) \, dv_0 \wedge d\phi_0 \wedge dv_1 \wedge d\phi_1,
$$

with

$$
(11) \quad 0 \leq \phi \leq \pi, \quad 0 \leq \phi_0, \phi_1 \leq 2\pi.
$$

By Euler's formula[11]

$$
(12) \quad \kappa_n^{(i)} = \kappa_1^{(i)} \cos^2 \phi_i + \kappa_2^{(i)} \sin^2 \phi_i,
$$

where $\kappa_1^{(i)}$, $\kappa_2^{(i)}$ are the principal curvature of ∂D_i, and

$$
(13) \quad \int_0^{2\pi} \sin^2 t \, dt = \int_0^{2\pi} \cos^2 t \, dt = \pi,
$$

we have

$$
(14) \quad \int (\kappa_n^{(i)})^2 dv_i \wedge d\phi_i
$$

$$
= 3\pi \int \left(\frac{\kappa_1^{(i)} + \kappa_2^{(i)}}{2} \right)^2 dv_i - \pi \int \kappa_1^{(i)} \kappa_2^{(i)} \, dv_i = \pi(3\tilde{H}_i^{(2)} - \tilde{K}_i),
$$

where K_i and H_i are, respectively, Gaussian curvature and mean curvature of ∂D_i. Putting (14) into (10) we proved (3).

The Proof of Theorem 2. Let the boundary ∂D of the domain D be a simple hypersurface of class C^2. It is known that at each point of a hypersurface Σ in $\mathbb{R}^n$ there are $n - 1$ principal directions and $n - 1$ principal curvatures $\kappa_1, \cdots, \kappa_{n-1}$. If $d\sigma$ denotes the area element of Σ, then the r-th integral of mean curvature

$$
(15) \quad M_r(\Sigma) = \binom{n-1}{r}^{-1} \int_\Sigma \{\kappa_{i_1}, \cdots, \kappa_{i_r}\} \, d\sigma,
$$

where $\{\kappa_{i_1}, \cdots, \kappa_{i_r}\}$ denotes the r-th elementary symmetric function of the principal curvatures. In particular, M_0 is the area and M_{n-1} is a numerical multiple of the degree of mapping of Σ into the unit hypersphere defined by the field of normals. Let D_0 and D_1 be two domains in $\mathbb{R}^n$ bounded by the simple hypersurfaces ∂D_0 and ∂D_1, which we assume to be of class C^2; here M_i^0, M_i^1

are the i-th integrals of mean curvature of ∂D_0 and ∂D_1, respectively. Chern's kinematic fundamental formula [1,2] is

$$
(16) \qquad \int_{\{g:D_0 \cap gD_1 \neq \emptyset\}} \chi(D_0 \cap gD_1)\, dg
$$

$$
= O_{n-1} \cdots O_2 [O_n(V_0 \chi(D_1) + V_1 \chi(D_0)) + \frac{1}{n} \sum_{h=0}^{n-2} \binom{n}{h+1} M_h^0 M_{n-2-h}^1],
$$

where dg is the kinematic density for $\mathbb{R}^n$, $\chi(\cdot)$ is the Euler-Poincaré characteristic. O_m is the volume of the unit sphere in $\mathbb{R}^m$, and its value is given by

$$
(17) \qquad O_m = \frac{2\pi^{\frac{m}{2}}}{\Gamma(\frac{m}{2})}.
$$

Let M^p, M^q be two piecewise smooth submanifolds in $\mathbb{R}^n$ with finite volumes $\sigma_p(M^p)$, $\sigma_q(M^q)$ and dimensional p, q, respectively. Suppose $p + q - n \geq 0$, then we have Santaló's formula [1]

$$
(18) \qquad \int_{\{g:M^p \cap gM^q \neq \emptyset\}} \sigma_{p+q-n}(M^p \cap gM^q)\, dg
$$

$$
= \frac{O_{n+1} O_n \cdots O_2 O_{p+q-n+1}}{O_{p+1} O_{q+1}} \sigma_p(M^p)\, \sigma_q(M^q),
$$

where $\sigma_m(M^m)$ is the volume of an m dimensional submanifold.

Now for our special case, $n = 3$, $p = q = 2$, i.e., $M^p = \partial D_0$ and $M^q = \partial D_1$, S. S. Chern's formula (16), Santaló's formula (18), respectively, read

$$
(19) \qquad \int_{\{g:D_0 \cap gD_1 \neq \emptyset\}} \chi(D_0 \cap gD_1)\, dg
$$

$$
= 8\pi^2(V_0 \chi(D_1) + V_1 \chi(D_0)) + 2\pi(F_0 \tilde{H}_1 + F_1 \tilde{H}_0),
$$

$$
(20) \qquad \int_{\{g:\partial D_0 \cap g\partial D_1 \neq \emptyset\}} L_{C_g}\, dg = 2\pi^3 F_0 F_1,
$$

where L_{C_g} is the volume of the intersection curve $C_g = \partial D_0 \cap g\partial D_1$, i.e., the arc length of C_g. The curve C_g may be composed of several components (i.e., each of those is a closed simple curve).

By Fenchel's theorem

$$
(21) \qquad \int_{C_i} \kappa_{C_i}\, ds \geq 2\pi,
$$

the equality holds when and only when curve C_i is a plane convex curve, and Hölder's inequality we have

$$(22) \quad 2\pi \le \int_{C_g} \kappa_{C_g} ds \le \left(\int_{C_g} 1^2 \cdot ds \right)^{\frac{1}{2}} \left(\int_{C_g} \kappa_{C_g}^2 ds \right)^{\frac{1}{2}} = (L_{C_g})^{\frac{1}{2}} \left(\int_{C_g} \kappa_{C_g}^2 ds \right)^{\frac{1}{2}},$$

i.e.,

$$(23) \quad 2\pi \le (L_{C_g})^{\frac{1}{2}} \left(\int_{C_g} \kappa_{C_g}^2 ds \right)^{\frac{1}{2}}.$$

Integrating (23) with respect to kinematic density dg and using Hölder's inequality again we get

$$(24) \quad 2\pi \int_{\{g:\partial D_0 \cap g\partial D_1 \neq \emptyset\}} dg \le \int_{\{g:\partial D_0 \cap g\partial D_1 \neq \emptyset\}} (L_{C_g})^{\frac{1}{2}} \left(\int_{C_g} \kappa_{C_g}^2 ds \right)^{\frac{1}{2}} dg$$

$$\le \left(\int_{\{g:\partial D_0 \cap g\partial D_1 \neq \emptyset\}} L_{C_g} dg \right)^{\frac{1}{2}} \left(\int_{\{g:\partial D_0 \cap g\partial D_1 \neq \emptyset\}} \left(\int_{C_g} \kappa_{C_g}^2 ds \right) dg \right)^{\frac{1}{2}}.$$

By (3), (20) and (24) we have

$$(25) \quad \int_{\{g:\partial D_0 \cap g\partial D_1 \neq \emptyset\}} dg \le \pi^2 \{ F_0 F_1 [(3\tilde{H}_0^{(2)} - \tilde{K}_0) F_1 + (3\tilde{H}_1^{(2)} - \tilde{K}_1) F_0] \}^{\frac{1}{2}}.$$

From

$$(26) \quad \int_{\{g:D_0 \cap gD_1 \neq \emptyset\}} \chi(D_0 \cap gD_1) \, dg \le N_0 \int_{\{g:D_0 \cap gD_1 \neq \emptyset\}} dg,$$

(19), (25) and (26) we get the kinematic measure

$$(27) \quad m\{g \in G : gD_1 \subseteq D_0 \quad \text{or} \quad gD_0 \subseteq D_1\}$$

$$= \int_{\{g:D_0 \cap gD_1 \neq \emptyset\}} dg - \int_{\{g:\partial D_0 \cap g\partial D_1 \neq \emptyset\}} dg$$

$$\ge \frac{1}{N_0} [8\pi^2 (V_0 \chi(D_1) + V_1 \chi(D_0)) + 2\pi (F_0 \tilde{H}_1 + F_1 \tilde{H}_0)]$$

$$- \pi^2 \{ F_0 F_1 [(3\tilde{H}_0^{(2)} - \tilde{K}_0) F_1 + (3\tilde{H}_1^{(2)} - \tilde{K}_1) F_0] \}^{\frac{1}{2}}.$$

4. Remark

If D_0 and D_1 are convex domains (in $\mathbb{R}^3$), we have $\chi(D_0) = \chi(D_1) = \chi(D \cap gD_1) = N_0 = 1$. Then Chern's fundamental formula (17) becomes Blaschke's formula, i.e.,

$$(28) \qquad \int\limits_{\{g:D_0 \cap gD_1 \neq \emptyset\}} dg = 8\pi^2(V_0 + V_1) + 2\pi(\tilde{H}_0 F_1 + \tilde{H}_1 F_0).$$

One of the results in [6] is an easy consequence of our Theorem 2.

THEOREM 3. *Let D_i $(i = 0, 1)$ be convex domains in 3-dimensional Euclidean space $\mathbb{R}^3$ with boundaries ∂D_i of class C^2. Assume that V_i, F_i, $\tilde{K}_i$, $\tilde{H}_i$ and $\tilde{H}_i^{(2)}$ are the volumes of D_i, the surface areas of D_i, the total Gaussian curvatures of ∂D_i, the total mean curvatures and the total square mean curvatures of ∂D_i, respectively. Then a sufficient condition for D_0 to contain, or to be contained in, D_1 is*

$$(29) \quad 8\pi(V_0 + V_1) + 2(F_0 \tilde{H}_1 + F_1 \tilde{H}_0)$$

$$- \pi\{F_0 F_1[(3\tilde{H}_0^{(2)} - \tilde{K}_0)F_1 + (3\tilde{H}_1^{(2)} - \tilde{K}_1)F_0]\}^{\frac{1}{2}} > 0.$$

Moreover,
(i) if $V_0 > V_1$ or $F_0 > F_1$, then D_1 can be contained in D_0;
(ii) if $V_1 > V_0$ or $F_1 > F_0$, then D_1 can contain D_0.

Using Gauss-Bonnet formula $\tilde{K}_i = 4\pi$, our formula (29) becomes

$$(30) \quad 8\pi(V_0 + V_1) + 2(F_0 \tilde{H}_1 + F_1 \tilde{H}_0)$$

$$- \pi\{F_0 F_1[3(\tilde{H}_0^{(2)} F_1 + \tilde{H}_1^{(2)} F_0) - 4\pi(F_0 + F_1)]\}^{\frac{1}{2}} > 0.$$

Of course, these conditions are not necessary.

References

1. L. A. Santaló, *Integral Geometry and Geometric Probability*, Addison-Wesley, Readings, Math. (1976).
2. S. S. Chern, *On the kinematic formula in the euclidean space of n dimensions*, Amer. J. Math. **74** (1952), 227-236.
3. Delin Ren, *Introduction to Integral Geometry*, Shanghai Press of Sciences and Technology (1987).
4. C-S. Chen, *On the kinematic formula of square of mean curvature*, Indiana Univ. Math. J. **22** (1972-73), 1163-1169.
5. Gaoyong Zhang, *A sufficient condition for one convex body containing another*, Chin. Ann. of Math. **9B(4)** (1988), 447-451.
6. Jiazu Zhou, *The generalizations of Hadwiger's theorem, the possibilities for a convex domain to fit another in $\mathbb{R}^3$*, submitted.
7. Jiazu Zhou, *Note on the sufficient condition for a convex domain to contain another in $\mathbb{R}^4$*, submitted.
8. Jiazu Zhou, *The analogues of Hadwiger's theorem in space $\mathbb{R}^n$, the sufficient condition for a convex domain to enclose another*, submitted.
9. P. R. Goodey, *Connectivity and free rolling convex bodies*, Mathematika **29** (1982), 249-259.
10. R. Howard, *The kinematic formula in riemannian geometry*, to appear as a Memoir of AMS.
11. Michael Spivak, *A Comprehensive Introduction to Differential Geometry (II)*, Publish or Perish, Inc. (1979).
12. Jiazu Zhou, *When can one domain enclose another in space*, submitted.
13. Jiazu Zhou, *The kinematic formula for the powers of mean curvature of hypersurfaces and the analogues of Hadwiger's theorem in $\mathbb{R}^n$*, submitted.
14. H. Hadwiger, *Genenseitige Bedeckbarkeit zweier Eibereiche und Isoperimetrie*, Viertejsch. Naturforsch. Gesellsch. Zürich **86** (1941), 152-156.
15. H. Hadwiger, *Überdeckung ebener Bereiche durch Kreise und Quadrate*, Comment. Math. Helv. **13** (1941), 195-200.
16. Yu. D. Burago & V. A. Zalgaller, *Geometric Inequalities*, Springer-Verlag Berlin Heidelberg (1988).

Current address: Department of Mathematics, Temple University, Philadelphia, PA 19122

E-mail address: Zhou@euclid.Math.Temple.edu

Recent Titles in This Series

(*Continued from the front of this publication*)

(See the AMS catalog for earlier titles)